CONTESTING THE CLIMATE UNTHINKABLE

Contesting the Climate Unthinkable

Latin American Cultural Responses to a Warming World

Edited by Azucena Castro,
Gianfranco Selgas, and Ken Benson

Foreword by Gabriel Giorgi

UNIVERSITY OF FLORIDA PRESS

Gainesville

Cover: Artisanal sluice in the installation *Lo que la mina te da, la mina te quita,* by Ana Alenso. Photo by Ana Alenso, 2020, Lo que la mina te da, la mina te quita.

This book will be made open access within three years of publication thanks to Path to Open, a program developed in partnership between JSTOR, the American Council of Learned Societies (ACLS), University of Michigan Press, and The University of North Carolina Press to bring about equitable access and impact for the entire scholarly community, including authors, researchers, libraries, and university presses around the world. Learn more at https://about.jstor.org/path-to-open/

Publication of this work made possible by a Sustaining the Humanities through the American Rescue Plan grant from the National Endowment for the Humanities.

Copyright 2025 by Azucena Castro, Gianfranco Selgas, and Ken Benson

Excerpts from chapter 6 were previously published in *Fictional Environments: Mimesis, Deforestation, and Development in Latin America* by Victoria Saramago, copyright © 2021 by Northwestern University Press. Published 2021. All rights reserved.

All rights reserved

Published in the United States of America

30 29 28 27 26 25   6 5 4 3 2 1

DOI: https://doi.org/10.5744/9781683405528

A record of cataloging-in-publication information is available from the Library of Congress.
ISBN 978-1-68340-552-8 (cloth)
ISBN 978-1-68340-557-3 (paper)

UF PRESS

University of Florida Press
2046 NE Waldo Road
Suite 2100
Gainesville, FL 32609
floridapress.org

GPSR EU Authorized Representative: Mare Nostrum Group B.V., Mauritskade 21D, 1091 GC Amsterdam, The Netherlands, gpsr@mare-nostrum.co.uk

A las tierras y gentes de nuestra América

CONTENTS

List of Figures ix
Foreword: Latin America and the "Unthinkable" xi
Acknowledgments xv

Introduction: Contesting the Climate Unthinkable; Latin America and the Warming World 1
Gianfranco Selgas, Azucena Castro, and Ken Benson

Part 1. Reimagining the Land: Amerindian Aesthetics and Alternative Ecologies for a Deranged World

1. Dying Well: Micropolitics of Life Against *Terricidio* in Mapuche Poetics 25
   Montserrat Madariaga-Caro

2. The Art of an Ecological Constitution: Beyond Derangement in Chile 41
   Paul R. Merchant

3. Entangled Ruins: Nature and Postcatastrophic Landscapes in Hiram Bingham's *Inca Land* and José de la Riva-Agüero's *Paisajes peruanos* 57
   Andrés Ernesto Obando

Part 2. The Political Ecology of Artistic Forms: Culture, Extractivism, and Environmental Imagination

4. What Is the Conflict About? Ecology, Ontology, and Decoloniality in Joseph Zárate's *Guerras del Interior* 75
   José Carlos Díaz Zanelli

5. Archives of the Planetary Mine: Art, Political Ecology, and Media Geology in Chile and Venezuela 92
   Gianfranco Selgas

6. Fiction Writing and Environmental Conservation in Latin America: The Cases of João Guimarães Rosa and Alejo Carpentier   111
Victoria Saramago

Part 3. Disrupting the Sensible Order in Landscapes of Contamination, Toxicity, and Wildfires

7. Embodying Anthropocene Awareness in *ecogótico rioplatense*   127
Allison Mackey

8. Haunting Trees in the Global South: Image and Life in the Rubble of Climate Change   144
Roberto Robalinho

9. Toxic Transits: Ghostly Double Gazes, Slow Violence, and North-South Ecologies of Inequalities in the Films *Sealed Cargo* (Bolivia) and *Arica* (Chile)   161
Azucena Castro

Part 4. Contesting Environmental Catastrophism: Queer Bodies, Afro-Indigenous Epistemologies, and the Crisis of Futurity

10. Culture, Inequality, and Queer Ecology in Rita Indiana's *La Mucama de Omicunlé*   183
Emily Baker

11. Toward a Cosmopolitics of the Image: Notes for a Possible Ecology of Cinematic Practices   201
Sebastian Wiedemann

12. Brazilian Afrofuturism, Climate Apocalypse, and Heuristic Function   216
Patrick Brock

Part 5. Ethnographic and Poetic Interventions

13. *Choike Pürrun:* The Mapuche People's Sacred Dance   237
Jasmin Belmar Shagulian

14. The Paths Not Seen: How I Structured My Representation of Nature   245
Igor Barreto

List of Contributors   257
Index   259

FIGURES

5.1. Industrial facility, screenshot of film *Desviar la inercia*   97
5.2. Chuquicamata, screenshot of film *Desviar la inercia*   98
5.3. Copper, screenshot of film *Desviar la inercia*   99
5.4. Artisanal sluice, *Lo que la mina te da, la mina te quita* installation   103
5.5. Mining cabin, *Lo que la mina te da, la mina te quita* installation   106
9.1. Aerial view of Arica town in Chile, film still, *Arica*   166
9.2. Aerial view of Boliden's headquarters in Sweden, film still, *Arica*   167
9.3. Sign showing Africa or South America as destinations of toxic waste, film still, *Sealed Cargo*   171
9.4. Train carrying the toxic cargo across the Altiplano, Bolivia, film still, *Sealed Cargo*   173
9.5. Indigenous defenders of the territory, film still, *Sealed Cargo*   175
11.1. Screenshot of film *Abyss* by Sebastian Wiedemann   203
11.2. Screenshot of film *Xapirimuu* by Sebastian Wiedemann   207
13.1. The choike's paws signaling the four cardinal points, image by Aiyana Shagulian   239
13.2. "Choike Lo Prado 2015" by Ramón Millache, video still   240
13.3. "Choike Pürrün familia Porma," video still   242

FOREWORD

Latin America and the "Unthinkable"

In 2013, early in the conversation about the Anthropocene, French historians Christophe Bonneuil and Jean-Baptiste Fressoz, in a well-known book, spoke of "l'évènement Anthropocène," translated into English as the "shock of the Anthropocene." They used the figure of the *évènement* or the "shock" to mark the Anthropocene as a new configuration of public debate about the environmental crisis and epochal mutation. A few years later, Amitav Ghosh published *The Great Derangement: Climate Change and the Unthinkable*, a highly influential intervention into the relation between literature (or literary narrative, to be more precise) and climate crisis, a crisis that, in his argument, needs to be brought back to narration from the shadows of the "unthinkable"—since the planetary has been, according to Ghosh, marked as the *unthought* by Western modernity. "Shock," "évènement," "unthinkable": the formulas reflect the affects mobilized during the initial debates and conceptualizations around the Anthropocene, as if the "irruption of Gaia" (to use yet another formula) and the emergence of the planetary could only be articulated under the sign of a sudden and radical disturbance. These affects are not, however, homogeneously shared across the world: the irruption of the planetary does not have a similar affective response independently from where we are situated. Evidently, the "shock of the Anthropocene" comes as less of a "shock" in some latitudes than in others—especially in Latin America, the very laboratory of colonial and extractive practices that subtends and makes possible modern capitalism. Indeed, the entanglement between historical and geological temporalities is hardly big news in regions shaped by colonialism and extractivism. Given the fact that Latin America has been a global laboratory of capitalist extraction for more than five hundred years, the entanglement between geological time and social time, between the human and the nonhuman, and between ancestrality and futurity has been a crucial feature in Latin American cultural traditions for a long time.

If our present feels the urge to discuss these issues due to the ongoing environmental crisis, the Latin American cultural archives are an embodiment of the *longue durée* of these issues.

It is, then, about time that these historicities, knowledges, and perspectives resituate the debate on the Anthropocene and the climate crisis. This is the key intervention carried out by this volume. *Contesting the Climate Unthinkable: Latin American Cultural Responses to a Warming World*, edited by Azucena Castro, Gianfranco Selgas, and Ken Benson, is a timely cartography of the many vectors Latin American cultures and archives bring to the center of the climate crisis debate.

Let's see, very briefly, what is at stake here. As it is well known, in *The Great Derangement* Amitav Ghosh argued that the modern novel configured, as a genre (i.e., in its very formal structure), the obstruction of environmental, planetary, more-than-human forces and therefore impeded its own ability to shed light on the ecological crisis—therefore, the novel is a crucial factor in the making of the *unthinkable* that is the climate crisis from the perspective of the literary archive. In his view, the modern novel—which for Ghosh is fundamentally the Western realist novel—elaborated narrative grammars that are too anthropocentric, too sociocentric (understanding the social as human-centered), and consequently severely limited its own ability to express and format nonhuman forces, perspectives, and scales that were reserved for less prestigious genres such as sci-fi or the fantastic (and, one may add, children's literature.) By doing so, Ghosh argues, the novel suffocated its own ability to understand the environmental mutation that subtends and makes possible the very emergence of modern literature itself.

One may wonder what would happen if Amitav Ghosh had in his literary cartographies a text like *La vorágine* (1924), by Colombian writer José Eustasio Rivera, where the extraction of rubber not only destroyed territories and communities under a vortex of extractivist violence but also mobilized more-than-human perspectives—mainly, that of the jungle—and conflicting temporal perspectives between human and nonhuman. Or a text like *Os sertões* (1905), by Euclides da Cunha, where the geological layers of the territory are active in ways that interfere and model the very civilizational project of the nation-state. The place of the nonhuman and planetary forces is quite central in these canonical artifacts. These are artifacts capable of changing arguments, periodization, and concepts since they articulate a historicity that the Global North had ignored (or repressed), and that now returns under the sign of the catastrophe and the "shock"—a configuration of historical experience in which the environmental, the more-than-human was never unconscious nor unthinkable. The very limits of the *thinkable* are, then, at stake when we decide what archives, what temporalities, what configurations of historical experience we mobilize, and with which we construct our critical tools.

Contesting the Climate Unthinkable offers an exceptional conceptual, aesthetic, and historical mapping of Latin American cultural archives that complicate, nuance, and challenge the conversation about the Anthropocene in the Anglosphere. These cultural responses are formal, conceptual, and political: they show the extent to which the environment has been at stake in a region shaped by multiple grammars of coloniality and extractivism. "Desediment the climate unthinkable" is the precise formula used by the editors, signaling the fact that the many cultural strata shaping Latin American imaginaries had always been actively thinking the environmental mutations brought by colonization, capitalist forces, and extractivist exploitation.

The chapters gathered by *Contesting the Climate Unthinkable* demonstrate the conceptual laboratory at work in Latin American environmental humanities. Among the many lines of research and discussion gathered here, I would like to highlight two vectors that, in many ways, underscore the crucial contributions coming from Latin American archives and critical practices. On the one hand, the insistent awareness about the entanglement between extractivist violence and climate change (in contraposition, for instance, to the focus on consumption as the main formula in ecological vocabularies in the North) finds in Latin America a key tradition, given the prominence of the multiple forms of extractivism in the making of Latin American nations. The planetary, the terraforming, the geological have been, then, already there in the cultural imaginaries from and about Latin America, precisely because its inscription in the global forces of capital has been, and still is, brutally characterized by the commodification of nature and the multiplication of the extractive frontier. *Contesting the Unthinkable* articulates the "reading machines" needed to activate this centrality of the planetary in Latin American cultural archives.

The other coordinate that emerges with singular richness and force from Latin America is the impact of indigeneity and ancestrality as political and aesthetic reconfigurations within cultural practices. *Contesting the Climate Unthinkable* offers, in this sense, an exceptional perspective on the ways Indigenous militancy, knowledges, and sensibilities impact upon a cultural tradition so forcefully shaped by Westernized forms epitomized by the idea of the *cultura letrada*, or lettered culture. We see an exponential increase not just in "visibility" of Indigenous interventions but fundamentally a persistent challenge to cultural forms (e.g., genres, enunciative positions, forms of cultural interpellations) coming from Indigenous intellectuals and artists. This is one of the most significant mutations in contemporary Latin American cultures, one that takes place not just in terms of "cultural diversity" but fundamentally in the territorial struggles against new waves of dispossession, violence, and neoextractivist forces. The political translation of this cultural mutation around indignity and its social resonance remains to be seen. But, in any case, going back to the "unthinkable" of the title, the volume reminds us that this cultural laboratory, these formal mutations, this critical "desedimenting" cannot take place if intellectual, cultural, religious, and political Indigenous practices would not have been active, even if they had not been at the center of artistic or political discussions. Again, the *longue durée* of these issues emerges under the light of the current disputes and debates.

Contesting the Climate Unthinkable offers an exceptional array of critical perspectives that mobilize, under the light of the present urgencies, the many archives and interventions coming from Latin America that reshape conversations, narratives, and knowledge production about the climate crisis. It shows the extent to which Latin America has been, from early on, the battleground of the colonial modeling of the planet that we now call the Anthropocene, as well as a critical laboratory from where other temporalities, other memories, and other ways of living can be thought, remembered, reactivated.

Gabriel Giorgi

ACKNOWLEDGMENTS

We are deeply grateful for the institutional support provided by Stockholm University, Stanford University, and University College London, where we have worked on this manuscript over several years. We also acknowledge the financial support of Stockholm University's Literary Studies Unit, which made this project possible, as well as postdoctoral funding from the Swedish Research Council and the British Academy.

We extend our sincere appreciation and gratitude to the colleagues who participated in the workshop, engaged in discussions around the project, and provided invaluable feedback. In particular, we thank Gisela Heffes, Victoria Saramago, Igor Barreto, Gabriel Giorgi, and Elizabeth Petinarolli for their insightful comments and engagement with this project at various stages. We are equally grateful to the scholars and researchers who contributed chapters to this volume—your commitment and engagement were essential in bringing *Contesting the Climate Unthinkable* to fruition.

We also wish to acknowledge the editorial support and meticulous work of Stephanye Hunter and the team at the University of Florida Press. We are thrilled that our project found such a fitting home. Our gratitude extends to the external reviewers, whose critical input and constructive feedback greatly improved the original manuscript.

Introduction

Contesting the Climate Unthinkable

Latin America and the Warming World

GIANFRANCO SELGAS, AZUCENA CASTRO, AND KEN BENSON

The Progress of This Storm, Andreas Malm's influential work in environmental debates, characterizes our current era as the "warming condition," a period where extreme weather patterns dominate our present and shape our future. This condition has evolved gradually, emerging from the systematic extraction of fossil fuels dating back hundreds of millions of years, coupled with the widespread combustion practices developed over the last two centuries. As Malm argues, the warming condition is both a historical and material phenomenon. Climate change and the accumulation of excess heat in the Earth's sediments and atmospheres result from the cumulative impact of emissions and the stacking of CO_2 over time. This environmental crisis is intertwined with the biocide that has emerged as historical patterns of organizing and exploiting nature, both human and nonhuman, continue to drive other forms of accumulation, particularly in terms of capital and power. According to Malm (11–12), however, despite the stark realities of the warming condition and the facts of global warming, culture remains willfully oblivious, perpetuating a cycle of everyday suppression and widespread denial. *Contesting the Climate Unthinkable* contends instead that culture—understood here as a broad assemblage of social and artistic practices, ideologies, and institutions—has been essential in reshaping conversations, narratives, and knowledge production about the climate crisis.

The warming condition primarily manifests among communities situated in the peripheries of the capitalist world economy, notably in the Global South. In these regions, a confluence of factors—including geographical vulnerability, economic dependence, historical colonial injustices, environmental degradation, and political marginalization—exacerbates the effects of climate change. In Latin America, for instance, the acceleration of the climate crisis has precipitated unprecedented environmental catastrophes. Recently, Uruguay declared a national emergency in 2023 due to water reserves depletion, while heatwaves and megadroughts in Argentina, Chile, Colombia, and Venezuela intensified wildfires in 2024. These instances underscore merely a fraction of the impact of capitalist-driven climate change on biodiversity, prompting ethical, cultural, and political inquiries about the future survival of human and nonhuman populations and ecosystems in the region and the planet. While recent cultural scholarship has pointed to a "crisis of culture, and thus of the imagination" (Ghosh 9) to account for the extreme climate events we are witnessing, Latin American cultural productions have been exploring ways to make these events *thinkable* and graspable, resorting to aesthetic modes that bridge the documentary register with the imaginative modes.

Cultural expressions such as films, novels, installations, photographs, videos, and performances can shed light on the layers of violence and destruction that underpin these extreme—previously unthinkable—events taking place in Latin America. The region's multimedia cultural production, as explored in this volume, often depicts toxic, burned, devastated, and ravaged environments. However, they also offer potential alternatives rooted in the region's cultures, geographies, and practices. *Contesting the Climate Unthinkable* recognizes the potential of cultural products and practices to illuminate the multiscalar dimensions of climate change amid extractivist violence. By signaling toward the many cultural strata shaping Latin American environmental imaginaries, the chapters gathered here "de-sediment" the entangled layers between nature, society, and extractivist violence (Rivera Garza 23), contesting the material and discursive dimensions of the climate unthinkable.

In this volume, we do not separate temporal aesthetics (future possibilities) from forensic aesthetics (of the body and earth sediments). As explored in each chapter, we believe Latin American realities acknowledge the importance of understanding these together, as realities and imaginaries go beyond any single metaphor and the division between territory and temporalities (Giorgi). That is why we provide a variety of perspectives to showcase the work of the contributors and creative endeavors in the region.

In the poetic intervention included in this volume, Venezuelan poet Igor Barreto suggests that poetry can illuminate "the paths not seen" and reveal the material and temporal lines of violence and destruction that connect

Latin America with the planet. While the multiscalar paths leading to socioecological emergencies, like the ones affecting the region right now, are complex and intertwined, cultural expressions can, as Barreto proposes, make graspable the unequal transformations of the environment and aid in envisioning more just futures. Latin American cultural products, as the ones studied in this volume, are attentive to what the character Benjamin Cordero in Barreto's dream advises the poet to do when writing a poem: "Do it in an unclean spirit, the dirtiest you can in the world." The films, novels, photographs, installations, and videos studied here embody such conditions as they excavate deep into the environmental violence to open the multiscalar denseness of climate change and enable perceptions into the lived experiences of environmental injustice. To put it differently, the environments registered in these cultural products are imbued in "a corrupted, invaded, impure representation: a *cordillera* [mountain range] of garbage," as Barreto's poem "Landscape" suggests. Such cultural immersion in the "unclean" has the potential to lay out alternatives rooted in the social and aesthetic practices of the region, thus contesting the unthinkable from the Global South.

At the turn of the century, Mary Louise Pratt (*Planetary Longings*) identifies a turn from the "post" and the "global" to the "planet" and "planetary." This is related to a "crisis of futurity" (Pratt 7) signaled by uncertainty about the yet-to-come in the face of the diverse geological and temporal scales that traverse today's socioenvironmental events. This planetary condition of bewilderment is generated by the multiscalar forces of present events that connect the geological to the cosmological and the microscopic in unprecedented ways. Latin American aesthetics and artistic works pay attention to the planetary turn by decentering the human through new arrangements of the sensible (Andermann, Giorgi, and Saramago 2–4). These arrangements reconfigure the sensorium that allows us to imagine, perceive, narrate, and think about the passage from the global to the planetary. Building up from this scholarship, *Contesting the Climate Unthinkable* brings to the fore situated creative and critical practices that not only show a local-planetary aesthetic passage but elaborate on forms of territorial resistance to the abstraction of what has been called "the climate unthinkable," or the ungraspable magnitude of the warming condition—we will come back to this in short. Within this planetary turn, we place Latin America along creative productions of the South. Creative works from the Global South contest the unthinkable by documenting the struggles of communities at the front line of environmental devastation and telluric trauma in imaginative ways, underscoring that thinking, writing, and creating from the Global South matters. The chapters included in this volume offer a window to hear the voices of affected communities and territories while also imagining alternative realities and speculating on more just futures.

Resisting and Imagining from the South

This volume engages with the ongoing debates on environmental representations in the region's cultural productions, and it explores how they grapple with the warming condition by engaging with the socioecological transformations endured in these geographies. This volume focuses particularly on artistic forms from the Hispanic Caribbean, the Southern Cone, the Andes, and the Amazon, between 1920 and 2022. It covers the heyday of Western economic developmentalism triggered by the "great acceleration," a paradigm that promised bright futures and progress based on an increased growth rate (Escobar, *La invención* 79; French and Heffes 125–27; Heffes, *Políticas de la destrucción*; Saramago, *Fictional Environments*), and it extends into the twenty-first-century commodities supercycle. Since the 1920s, the sustained rise in the prices of Latin American raw materials allowed for different stages of economic growth in the region. Historically, this made natural resources the economic engine and the mainstay of regional policies, positioning extractivism as an integral part of the process of capital accumulation relevant to this day (Svampa). Such a process, rooted in the consolidation of a fossil fuel–driven modernity (Szeman and Boyer 1–2), has drastically reconfigured the geopolitics of extractivism, economic and racial inequalities, as well as unjust distribution of climate change events in almost every country in the region. By doing so, this process exacerbated the effects of climate change in Latin America.

Fleshing out the entanglements between socioecological issues and the various forms of resource extraction as a world-ecological system within the Capitalocene (Moore, *Capitalism*), this volume sets out to address and imagine what Amitav Ghosh defined as "the unthinkable." In his influential *The Great Derangement: Climate Change and the Unthinkable*, Ghosh defines this notion as "the uncanniness and improbability, the magnitude and interconnectedness of the transformations that are now underway" (73). While Ghosh states that the unthinkable constitutes a crisis of culture and thus of the imagination (9)—we develop our approach to the unthinkable in the following section—the contributions gathered in this volume contest that statement by opening pathways of possibilities to thinking of/with damaged ecologies through practices of care, epistemological displacement, imagination, and horizontality. Questioning Ghosh's rationale, Mark Bould (*The Anthropocene Unconscious*) argues that climate change was an unconscious but pervasive topic in Western literature across genres that do not necessarily deal with environmental issues. *Contesting the Climate Unthinkable* relates to Bould's contention in that it also gathers some examples of cultural materials and expressions that do not show an explicit environmental concern. However, as the chapters presented in this volume will show, Latin American cultural expressions are both implicitly and explicitly concerned with the representation and documentation of diffused and differential geographies where ecological ravage is too tangible to remain solely as an unconscious manifestation.

Aesthetics play a key role in this effort, participating in organizing and regulating what is perceivable, sayable, and doable within cultural and political contexts. This is what Jacques Rancière calls the "distribution of the sensible," or the "delimitation of spaces and times, of the visible and the invisible, of speech and noise, which simultaneously determines the place and stakes of politics as a form of experience" (*Politics of Aesthetics* 13). These distributions shape perceptions and reinforce ideologies. They uphold existing hierarchies, inequalities, and systems of exclusion by determining who can participate in public discourse, who is considered worthy of attention and recognition, and whose voices are marginalized or silenced. However, these distributions are not merely passive reflections of existing power structures. Rancière points out that aesthetic practices, such as art, literature, and media, insofar as they actively reproduce and reinforce power and ideological structures, can also offer "a different sensorium, a different way of linking a power of affective perception and a power of signification" (*Politics of Literature* 14). The potential to enact forms of dissensus or to configure new regimes of thinkability—we will come back to this—implies that cultural production and practices disrupt and challenge existing distributions of the sensible, opening up new possibilities for political action and collective emancipation in the context of unthinkable climate change and the warming condition.

The chapters gathered in this volume will show how Latin American cultural production and practices have historically configured different regimes of thinkability that bring to the fore the problem of seeing and representing the temporalities and geographies of environmental degradation. As Rob Nixon puts it, this problem slips away at first glance due to its naturalization through forms of violence that are gradual, incremental, and often invisible, occurring over extended periods of time and across vast spatial scales. Nixon conceptualizes this as "slow violence" (2), or the violence that unfolds slowly and insidiously, escaping public attention and eluding conventional modes of representation. *Contesting the Climate Unthinkable* also makes the case for cultural production and artistic expression as an aid in developing ecological epistemologies attuned to the situatedness of local experiences of injustice in a planetary setting. The purpose of this volume is to problematize the cultural and imaginative crisis Ghosh pointed out and to foreground the forms that the warming condition takes in the region: What are the regimes of thinkability that can be derived from ways of reading and discussing the representations of the environment, as well as the making and unmaking of "nature" through different texts and media? And how do the human-nonhuman entanglements depicted in modern and contemporary cultural media articulate regimes of thinkability that flesh out local and dissident environmental epistemologies and expose racial, gender, and socioeconomic inequalities?

The drought in Uruguay, which left millions without access to safe drinking water, highlights the growing challenges posed by climate change, extreme weather events, biodiversity loss, and ecological tipping points—issues that were once unimaginable. These events take place across the territories and geographies of the continent, impacting human and nonhuman bodies that cohabitate in the region, and they should be approached with radical imaginaries that can aid us in thinking the world otherwise. As Jason W. Moore points out, "If your cosmology is Man against Nature, the political resolution of the climate crisis as a geohistorical event is unthinkable" ("There Is No Such Thing" 5). Hence, if we don't assume a radical position regarding our appraisal of socioecological relations and the imagination of a collective future, the scale and scope of human-made changes to the living world will continue to worsen.

Traversed by marginal and peripheral visions of society and the environment, Indigenous cosmologies and Amerindian thought, for example, can give us an alternative and radical solution to contest climate change from the South while encouraging us to question existing power structures and explore new possibilities for sustainable and just coexistence. They weave ecological knowledge and spiritual connections with nature into narratives that underscore the significance of living in harmony with the environment (Duchesne Winter). Similarly, as chapters gathered in Part 1, "Reimagining the Land: Amerindian Aesthetics and Alternative Ecologies for a Deranged World," will show, modern and contemporary Latin American authors, filmmakers, artists, and activists have increasingly turned their attention to environmental challenges, exploring the impacts of deforestation, industrial development, and resource extraction on local communities and ecosystems. Their cultural expressions have also become a relevant platform to convey political, ecological, and environmental justice/injustice concerns related to how race, class, gender, and other social identities intersect with environmental issues, leading to disproportionate exposure to pollution, lack of access to resources, and limited participation in decision-making processes. Here, climate change, environmental catastrophes, and socioecological crises emerge as tangible and real experiences. As climate change accelerates, its unthinkability encompasses alarming scenarios once regarded as improbable or distant.

Even though *Contesting the Climate Unthinkable* brings to the fore a novel way of approaching the region's cultural expressions by tackling the explicit and implicit representation of the unthinkable, this effort is built upon scholarship bridging cultural studies, the environmental humanities, political ecology, and environmental justice. Besides Ghosh's seminal study, postcolonial environmental humanities scholars such as Elizabeth DeLoughrey, Jill Didur, and Anthony Carrigan (*Global Ecologies*) have laid the theoretical and methodological ground floor to explore how the history of globalization and imperialism epitomizes a special challenge to the representation of environmental issues in regions from Africa and Asia to Oceania, South America, and North America. While *Contesting the Climate Unthinkable* builds upon this global and postcolonial environmental approach, it expands this perspective by unearthing situated, alternative methodological, and theoretical readings of the historical unfolding of socioecological transformations. By focusing on specific Latin American geographies (e.g., the Amazon Basin, the Bolivian Altiplano, the Andes, or the Caribbean Basin), the present volume enriches this postcolonial lens by opening new critical venues to understand the complex relations between culture and environmental transformations that have traversed their history as a Global South region. Similarly, Ilka Kressner, Ana María Mutis, and Elizabeth Pettinaroli (*Ecofictions*) employ Nixon's concept of slow violence to highlight the modes by which Latin American and Latinx cultures engage with ecological violence. The attempt to conceptualize, visualize, and document how ecological violence resists cultural, artistic, and political representation is also at the core of *Contesting the Climate Unthinkable*. Nonetheless, the contributions in the present volume expand this topic by emphasizing the creative and political possibilities of multimedia cultural production to grasp the scale and violence of climate change. Bringing to the fore Brazilian Afrofuturism, Caribbean and Central American queer ecologies, or Andean regionalisms is part of the effort to surface submerged perspectives to approach ecological violence and the complexities of the environmental crisis.

The latter resonates deeply with a body of critical literature that has been developing decolonial and radical approaches to Latin American cultural expressions in relation to climate change and the impact of extractivism in human and nonhuman communities. In her work, Macarena Gómez-Barris (xiii–xx) employs queer-femme and decolonial methods that question the capitalist gaze emerging in extractive zones coinciding with Indigenous lands exploited in the name of profit. Inspired by this method, *Contesting the Climate Unthinkable* gathers below-the-surface approaches that foreground local Indigenous and non-Indigenous communities that have been and continue to be affected by capitalist exploitation and the effects of climate change. This is also important regarding more-than-human representations in artistic expressions. Paul Merchant and Lucy Bollington (*Latin American*) have laid out the theoretical basis of a posthuman approach to study how Latin American visual culture challenges the meanings of the human and the ways by which humans and nonhumans shape one another. As this approach aims to decenter the Anthropos and foreground nonhuman and environmental materialities in Latin America, *Contesting the Climate Unthinkable* expands this scholarship by deploying less explored, peripheral approaches from Indigenous environmental law or *rioplatense* ecogothic, which not only show the limits of the human but illuminate the socioecological entanglements taking place in times of unprecedented climate change.

Over the past decades, cultural production in the region has provocatively engaged with the complexities of socioenvironmental issues, shedding light on the interplay between human societies, temporalities, and ecological contexts (Anderson and Bora, *Ecological Crisis*; Barbas-Rhoden, *Ecological Imaginations*; French and Heffes, *The Ecocultural Reader*; Fornoff, *Subjunctive Aesthetics*; Heffes, *Políticas de la destrucción*). In that regard, political ecology and environmental justice, understood as interdisciplinary fields that actively engage in cultural and social issues, have emerged as a powerful lens to understand the intricate connections between social, political, economic, and ecological processes. This volume brings together imagination, political ecology, and environmental justice by urging us to question mainstream climate narratives that tend to overlook the disproportionate burden borne by marginalized groups. The implications of climate change on food security, access to clean water, and livelihoods amplify social inequities, making the unthinkable scenarios of climate-induced poverty and displacement a harsh reality for many communities. In other words, this book makes a case for the modes by which Latin American multimedia cultural production unearths "sites of situated hope"—that is, partial and located knowledge that forges interactions between humans, nonhumans, and the world through solidarity in politics and shared conversations in epistemology (Castro and Selgas 592).

Similarly, situated environmental conflicts open disputes between the defense of life, profit, and the devastation of the territory, between Indigenous territorial rights and the extractive frontier. In these scenarios, new formats of collective action and artivism unfold, questioning different conceptions of democracy and its meaning for the defense of the commons (Merlinsky and Serafini 14). Grassroots activism and artivism have historically coalesced in the region to configure politics and poetics of resistance (Merlinsky and Serafini 20–21). Chapters in this volume explore environmental films addressing toxic waste dumping in Chile and Bolivia, activist expressions within Chilean social movements advocating for a new constitution, and Mapuche poetry and performances resisting terricide. Together, these works highlight embodied contestations and demonstrate how cultural and political actions in public spaces are crucial for imagining new configurations of the sensible. These also inscribe alternative narratives relevant to sustain the organization of anti-extractivist and anticolonial movements, water rights activism, community-based conservationism, and environmental justice at large. In that context, art becomes another form of expression and platform to channel collective action. Through its aesthetic and political potential, it activates the imagination and contests the unthinkable. To put it in Gabriela Merlinsky and Paula Serafini's terms: "In the context of activism, art has the ability to create a rupture in the everyday through the construction of immersive situations that disrupt the dynamics and rules of spaces" (20; our translation). These ruptures and distributions of the sensible allow us to contest the concepts that support the abstract basis for the climate unthinkable and the warming condition, as well as to visualize and experience other ontologies and ways of knowing and relating to the territory.

Understanding the Climate Unthinkable

The "unthinkable" has become a keyword in global environmental scholarship, connecting diverse interdisciplinary fields. Donna Haraway employs the "unthinkable" as a crucial concept: "Seriously unthinkable. Not able to think with" (30). The use of the term in Haraway refers to alternative forms of thinking beyond the rational argumentation of technocratic discourse, calling attention to the need to create new narratives that acknowledge the equality of all life forms available to "think with." Furthermore, Rosi Braidotti (*Posthuman Knowledge*) proposes that a posthuman framework can make concepts thinkable in a critical dimension where it becomes possible to analyze power relations in a relational manner. As Braidotti suggests, the unthinkable can be made thinkable in material and discursive chunks that activate other modes of experiencing human-nonhuman relations and generating knowledge about the world. In the Latin Americanist context, recent work by Carolyn Fornoff (*Subjunctive Aesthetics*) addresses how the subjunctive, hypothetical mode is explored in a series of works in Mexican cultural production to grapple with the question of how to make climate change and extractivism thinkable: "The subjunctive thinks something as thinkable" (3). This volume is inspired by this ideal and hypothetical mode in culture but it extends this approach by addressing the climate unthinkable in a series of cultural products from the region where the documentary mode is entangled with the speculative and hypothetical registers.

One of the challenges to the thinkability of today's climate events and the deep footprint of capitalism on the planet concerns how to apprehend the scalar relation between locality and the planetary. Gayatri Spivak (102) pioneered the idea of "planetarity" as a departure from the totalizing paradigm of modern globalization and from the concept of cosmopolitan globality as a way of constituting a vast world of unified and centralized references. Faced with the socioecological challenges of the turn of the century that account for crosses between culture and geology, the discourse of planetarity and its cosmocentric vision becomes a crucial exploration method to think from relational ethics of the situatedness of local experiences in a planetary setting. Amy Elias and Christian Moraru (xii–xxiii) discuss to what extent terms such as "planet" and "planetarity" constitute terminological substitutes for the idea of cosmopolitanism as a way of knowing and understanding the world, evaluating it, and taking perspectives on it. Planetarity, however, goes beyond hermeneutics and has to do with ontologies and material relationalities beyond the human perspective. Hence, planetarity enables us to think of the Earth as material, as a nonnegotiable ground for human and nonhuman life; the Earth's societies are conceived as a diverse and relational assemblage composed of humans and nonhumans. In this context, cultural practices interact with the materialities of the planet and the forms of life it hosts: "The planet as a living organism, as a shared ecology, and as an incrementally integrated system both embracing and rechanneling the currents of modernity is the axial dimension in which writers and artists perceive themselves, their histories, and their aesthetic practices" (Elias and Moraru xii).

In addition to this view, Héctor Alimonda ("La naturaleza colonizada") has pointed out the relevance of understanding the persistent coloniality that affects Latin American environments. Both as a biophysical reality (i.e., its flora, fauna, human inhabitants, the biodiversity of its ecosystems) and its territorial configuration (i.e., the sociocultural dynamics that coexist as part of these ecosystems), the region has been represented in global hegemonic thought as a marginal space that can be exploited, racialized, and reconfigured according to the needs of the current accumulation regimes. The colonial legacies, land tenure systems, and extractive economies have all played pivotal roles in shaping patterns of unthinkability and environmental exploitation in the region (Alimonda; Delgado Ramos; Leff). As chapters gathered in Part 2, "The Political Ecology of Artistic Forms: Culture, Extractivism, and Environmental Imagination," will show, cultural production that deals directly and indirectly with these issues functions as a form of political ecology that questions, in an experimental, imaginative, and radical way of thinking, the patterns of colonization, appropriation, and exploitation in the web of life. In other words, cultural expressions serve as an indispensable framework for comprehending what has been framed as the unthinkable and the intricate interactions between humans and the environment within the context of power dynamics and sociopolitical structures (Robbins xvi–xvii).

In a similar fashion, as Pratt (*Planetary Longings*) has suggested, the "unthinkable" in imperial history had often underpinned how the West kept hidden and out of sight its dynamics of conquering, othering, pillaging, extracting, and appropriating the Global South. Understanding and exposing such dynamics meant a revolution in perspective, which allowed us to apprehend how the history and geography of the region were imbued with such violence (Pratt 15). The latter is showcased in Part 3, "Disrupting the Sensible Order in Landscapes of Contamination, Toxicity, and Wildfires," which proposes that cinema, photography, and literature of the Global South have the potential to grapple with the unthinkability of contaminated, intoxicated, and burned environments by attending to relations between local and planetary scales.

As we have discussed so far, the unthinkable is a common denominator of Anthropocene discourses. Quoting Crutzen, Ghosh, Braidotti, and Haraway, Ramona Mosse contends that "the unthinkable is a recurring feature in theories of the Anthropocene, connected both to the futurity of scientific projections and to the fatal flaws of the economic and political systems that enable sustained exploitation of the earth's environments" (106). At the same time, we witness the emergence of what Timothy Morton (*The Ecological Thought*) has called "hyperobjects" (e.g., toxicity, plastics, radiation, global warming), understood as vast entities that transcend localities and temporalities beyond human lives and history. While hyperobjects "exist on almost unthinkable timescales" (19), in Latin America, as we mentioned before, such vast entities are entangled with histories of colonialism, imperialism, capitalism, and extractivism that have pillaged the environment in the name of progress and suppressed Afro-Indigenous knowledges.

From the field of Indigenous studies, Marisol de la Cadena ("The Production" 14) refers to the "unthinkable" as the absence of possibility to think with/ of Indigenous cosmologies and epistemologies from the Western colonizers' viewpoint, which delegitimized other bodies of knowledge and ontologies. Thus, approaching hyperobjects in Latin America requires "relational thinking" (de la Cadena and Blaser, *A World of Many Worlds*) as an ethical dimension to grasp the entanglements between the tangible and intangible worlds in the region's extractive histories and geographies. Chapters gathered in Part 4, "Contesting Environmental Catastrophism: Queer Bodies, Afro-Indigenous Epistemologies, and the Crisis of Futurity," speak to the latter as they foreground relationalities inspired by queer bodies and Afro-Indigenous cosmologies in science fiction and experimental philosophy. The works discussed in this section mobilize alternative epistemologies beyond dominant narratives that end up in environmental catastrophism. By doing so, the multimedia cultural products analyzed in these chapters bring to the fore the racial, gender, and socioeconomic inequalities rooted in the crisis of futurity in the region.

Central to Indigenous Latin American contributions to contesting the climate unthinkable is "buen vivir," which encompasses more than just a preservation of ancestral practices; it represents a profound reevaluation of modernity and unthinkability itself. The rapid hemispheric expansion of Indigenous ecosocial formulations reflects a deliberate effort to merge Indigenous knowledge with democratic practices and global environmental justice to reframe the way societies coexist with their surrounding environments. Buen vivir, often articulated through culture-specific Indigenous terms such as Sumak kawsay in Ecuador or Suma qamaña in Bolivia, embodies a vision of well-being that prioritizes interconnectedness, reciprocity, and respect for more-than-human beings and territories (Gudynas). But it also rejects the dominant paradigms of endless economic growth and massive exploitation of nature as a resource, offering instead another regime of thinkability rooted in ecosocial practices. As Davi Kopenawa puts it in *The Falling Sky: Words of a Yanomami Shaman,* these practices are rooted in the understanding of *nature* as animate, sentient, and multiple. By acknowledging more-than-humans as living organisms and beings, notions of *nature* as inert and purely instrumental matter are challenged, reframing their scale and significance, thus aiding us in thinking about what has been deemed unthinkable.

Alternative Indigenous epistemologies from the Andes, to name but a few examples, are covered by chapters contributed to this volume—for example, Mapuche land's justice ethos or AzMapu and holistic connections between humans and the cosmos or Choike Pürrun. They are also political interventions that represent an ecosocial gesture aimed at bridging the gap between Indigenous worldviews and mainstream discourses on ecology and democracy (Escobar, *Encountering Development;* de la Cadena, *Earth Beings*). Their political and existential stance, however, lies in how different artistic expressions and poetics influenced *by* or produced *within* the framework of Indigenous epistemologies stand explicitly for a more equitable and environmentally conscious planet. They seek to redefine the discourse on climate change and environmental justice in ways that are more compatible with ancestral cultural values and historically rooted experiences attuned with acute socioenvironmental emergencies—for example, settler colonialism, land grabbing and displacement, sustained deforestation, and human and nonhuman exploitation underwriting the social and natural history of many Indigenous communities in the region. These crises are neither remote nor abstract; they have been present for a long time and are unfolding as we speak.

Taken all together, the chapters of this volume contest the unthinkable of our warming condition by adding diverse dimensions of thinkability to today's socioecological crisis, foregrounding nonhuman afterlives in wildfires, local forms of environmental artivism, practices of care, decentralized epistemologies, and imaginations otherwise. By shedding light on these issues, *Contesting the Climate Unthinkable* charts the modes by which cultural productions pave the way for transformative and radical change as well as the pursuit of environmental justice in the region.

Chapters

Part 1, "Reimagining the Land: Amerindian Aesthetics and Alternative Ecologies for a Deranged World," explores the intersections between textual landscaping, visual art, and Indigenous poetics in the face of environmental devastation and the nature-humanity dichotomy.

Montserrat Madariaga-Caro's "Dying Well: Micropolitics of Life Against *Terricidio* in Mapuche Poetics" examines Mapuche poet, musician, and plastic artist Faumelisa Manquepillán's performance in the video *Wetruwe Mapuche*, focusing on how her work addresses the notion of *terricidio*: the murder of Mother Earth. The analysis explores the intersection of environmental issues, settler colonialism, and Indigenous resistance within Mapuche culture. It highlights the decentering of anthropocentric paradigms and reclaiming Mapuche language and cultural practices. In her analysis, Madariaga-Caro argues that Mapuche poetics offer a form of micropolitics that challenge the dominant anthropocentric model and advocate for justice in the face of terricidio. By embodying the AzMapu—the Mapuche land's justice ethos—Manquepillán creates an imaginative space that coexists with the oppressive colonial order and works toward the renewal of conditions necessary for Mapuche life and just livelihood. Overall, the chapter emphasizes the role of Mapuche poetics in addressing the climate crisis, racialized and gendered inequality, and asserting Indigenous rights and sovereignty.

Paul R. Merchant's "The Art of an Ecological Constitution: Beyond Derangement in Chile" discusses performances, art interventions, and films that emerged during the social movement in Chile toward a new constitution in 2022. The proposed new constitution in Chile included a chapter dedicated to nature and the environment, recognizing the importance of protecting biodiversity and the rights of nature. Engaging critically with Bruno Latour's and Kerry Whiteside's proposals concerning the representation of nature in constitutional discourses, Merchant analyzes how art and artivism question hegemonic modes of aesthetic representation in the context of the ecological crisis. These artistic works can be viewed as a call for a new, ecological constitutional settlement that may yet be answered. The archipelagic nature of Chile is also explored as a conceptual tool for understanding socioecological relationships. The chapter argues that artistic interventions can take an active part in imagining a more inclusive constitutional framework to embrace multiple values for nature.

Andrés E. Obando's "Entangled Ruins: Nature and Postcatastrophic Landscapes in Hiram Bingham's *Inca Land* and José de la Riva-Agüero's *Paisajes peruanos*" explores how modern narratives represent nature reclaiming ruins, reflecting anxieties and hopes tied to environmental catastrophes. While contemporary stories often use a "disaster frame" to depict climate change, some envision a future where nature thrives amid human ruins, blending past and future ruins. The chapter analyzes Hiram Bingham's *Inca Land* and

José de la Riva-Agüero's *Paisajes peruanos*, highlighting how these works depict the Peruvian Highlands through lush nature intertwining with ancient ruins. This depiction served to externalize modernity's fears by relocating them to its margins. Obando connects these historical views with current postapocalyptic visions, showing how colonial-era conceptualizations of nature continue to shape our imaginations of the future, ultimately suggesting that revisiting these ghosts of the past can clarify our present challenges.

Part 2, "The Political Ecology of Artistic Forms: Culture, Extractivism, and Environmental Imagination," addresses struggles against extractivism, the political geology of mining and capitalism, and the entanglements between environmental representation and conservation through chronicles, novels, film, and art installations.

José Carlos Díaz Zanelli's "What Is the Conflict About? Ecology, Ontology, and Decoloniality in Joseph Zárate's *Guerras del Interior*" interrogates culture's limits to consider the ecological conflicts involving Andean and Amazonian communities in Peru. Through the analysis of Joseph Zárate's *Guerras del Interior* (2018), a compilation of chronicles about the anti-extractivist struggle incarnated by three Peruvian Indigenous ecologist activists, this work sheds light on the debate about what is the issue at stake in these conflicts by examining the cases of the Andean peasant Máxima Acuña and the Amazonian activist Edwin Chota. This study exposes a collision of ontologies rather than of cultures, so dealing with them from a culturalist approach would lead to the systematic impasse that originates the structural violence portrayed in Zárate's chronicles. In addition, it exposes how the Indigenous ecologist mobilities depicted in these chronicles constitute a vivid and ongoing mode of resistance against the different forms of coloniality entailed by extractivism.

Gianfranco Selgas's "Archives of the Planetary Mine: Art, Political Ecology, and Media Geology in Chile and Venezuela" explores the intricate relationships between geology, politics, and culture through contemporary Latin American art, focusing on the socioecological impacts of mining. The discussion centers on Ana Alenso's works, including her film *Desviar la inercia* (2019) and her multimedia installation *Lo que la mina te da, la mina te quita* (2020). These works symbolize and materialize the entanglements of extractive capitalism, particularly in Chile and Venezuela. By examining the intersections of geology and political ecology, this study highlights how artistic practices can reveal and critique the socioeconomic and environmental consequences of mineral extraction. The concept of "archives of the planetary mine" is introduced to describe how art acts as a cultural archive, recording the dialectics of socioecological entanglements driven by mining. This approach situates Alenso's art within broader debates on the Capitalocene, political geology, and the commodification of nature, emphasizing the role of art in addressing and challenging environmental and social crises.

Victoria Saramago's "Fiction Writing and Environmental Conservation in Latin America: The Cases of João Guimarães Rosa and Alejo Carpentier" traces the relationship between fiction writing and environmental conservation in Brazilian author João Guimarães Rosa's novel *Grande sertão: Veredas* (1956) and Cuban author Alejo Carpentier's novel *Los pasos perdidos* (1953) and their impact on the creation and perception of national parks. The chapter argues that the symbolic significance attached to the settings of these novels (i.e., the Grande Sertão Veredas National Park in Brazil and Canaima National Park in Venezuela) inspired conservation initiatives. These initiatives sometimes involved real-world interventions that aimed to imitate the fictional landscapes described in the novels. Saramago's analysis shows how narrative conservation and conservationist narratives complement each other, surfacing another way of thinking about environmental representation. By emphasizing the reciprocal and contrastive relationship between conservation and fictionality, the chapter demonstrates how fictional works justify the objectives of national parks, legitimize their existence, and increase their visibility.

Part 3, "Disrupting the Sensible Order in Landscapes of Contamination, Toxicity, and Wildfires," addresses photography, film, and literature that seek to render visible the material traces of environmental destruction in bodies and territories. It contests the hidden logic of socioenvironmental injustices and proposes new forms of care and solidarity.

Roberto Robalinho's "Haunting Trees in the Global South: Image and Life in the Rubble of Climate Change" discusses the complex relations between image-making and Anthropocene by looking at the photographic portrayal taken in the aftermath of environmental disasters: the collapse of a mining dam in Brazil in 2015, the destruction of a sacred ghost gum tree in Northern Australia in 2013, and the Australian wildfires of 2019/20 known as the Black Summer. These images incorporate the complex layering of temporalities of the Anthropocene and simultaneously reveal a body and a hidden subjectivity that modernity has neglected. The study shows how the images depict a "nature composed of many bodies and different ontologies," thus rendering the spirit and body of these trees visible but in a disquieting ghostly manner. More than illustrating the destruction of the catastrophes, the images disrupt a sensible order by bringing to the fore alternative worlds and ontologies that have survived despite ongoing anthropogenic violence.

Azucena Castro's "Toxic Transits: Ghostly Double Gazes, Slow Violence, and North-South Ecologies of Inequalities in the Films *Sealed Cargo* (Bolivia) and *Arica* (Chile)" addresses the links between global toxic waste dumping in Global South geographies with environmental racism and colonialism, through the lens of the fiction movie *Carga Sellada* (2015) by Julia Vargas-Weise and the documentary *Arica* (2020) by Lars Edman and William Johansson Kalén. The chapter analyzes how the cinematic portrayal of the dumping of toxic waste generated in the Global North into the Bolivian Highlands and Arica in the Atacama region in Chile disquiets the hidden logic of global toxic transits. Castro argues that the toxic transits presented in these movies render visible the North-South ecologies of inequalities through "ghostly double gazes" that contribute to negotiating relations between transit and emplacement in the processes of global waste management. The chapter explores the potential of environmental films to reveal the slow violence (Nixon) of toxic waste dumping and allows for the expression of the toxic bios of affected communities.

Allison Mackey's "Embodying Anthropocene Awareness in *ecogótico rioplatense*" addresses how "the disastrous planetary consequences of our species and the violent rule of sovereign Anthropos" (Braidotti 10) are thought about in Río de la Plata gothic. This chapter examines recent engagements with gothic tropes (e.g., confinement, monstrosity, irrational fear, and the uncanny) by writers on either side of the Río de la Plata—namely, the short horror story *Bajo el agua negra* (2016) by Argentine writer Mariana Enríquez, and the dystopian novel *Mugre rosa* (2020) by Uruguayan writer Fernanda Trías. The figures of monstrous children, toxic rivers, and mutated bodies attend to historical and contemporary social injustices and modes of production specific to the region. The chapter suggests that far from awakening a horror that overwhelms and immobilizes the reader, these ambiguous and open-ended visions disrupt the division between normal and abnormal and enable us to consider the possibility of a multispecies ethics of care.

Part 4, "Contesting Environmental Catastrophism: Queer Bodies, Afro-Indigenous Epistemologies, and the Crisis of Futurity," gathers chapters that critically address environmental catastrophism. It foregrounds queer bodies and Afro-Indigenous epistemologies in science fiction and experimental philosophy as alternatives to the dominant narratives that can only imagine the future under intensified extractive capitalism, devoid of communal solidarity.

Emily Baker's "Culture, Inequality, and Queer Ecology in Rita Indiana's *La Mucama de Omicunlé*" analyzes Rita Indiana's novel *La Mucama de Omicunlé* (2015), a not-so-futuristic tale of a society coping with a deadly virus while facing climate emergency and severe biodiversity loss. The novel raises ethical questions about escapism, self-actualization, inequality, racism, homophobia, and the upholding of dualisms of sex, class, and nature. Bringing together queer, ecological, and Marxist theory, Baker argues that the novel displays a self-reflexive focus on art, at once a means of changing the way we view the world and a perpetrator of inequality in the context of social and environmental issues. Baker shows how this achieves the type of ecocritique advocated by Timothy Morton, which involves examining the ecological aspects of literature. The chapter demonstrates that an ecological Marxist lens is valuable for understanding the underlying critique of racialized and gendered inequality within the novel, advocating for a multiclass, multiracial, multigender, and anticapitalist movement.

Sebastian Wiedemann's speculative chapter, "Toward a Cosmopolitics of the Image: Notes for a Possible Ecology of Cinematic Practices," offers a philosophical fabulation exploring a "cosmopolitics of the image" by combining Yanomami cosmologies of the Brazilian Amazon, more-than-human philosophy, and autoethnographic reflections. The chapter focuses on how these encounters between worlds unfold through experimental audiovisual works that move away from binary thinking to expose perceptual and material bifurcations of nature for a cosmopolitical cinema to come. The ecologies of cinematic practices explored in this chapter put forward a conception of otherness as the proliferation of differences that have the potential to challenge the image-object dichotomy. This experimental chapter explores images as matter in ideas about "cinema by other means," "nonhuman filmmakers," and "shamanic cinema" that make up interstitial modes of existence as *imargins*—image-margins—rendered by cinema.

Patrick Brock's "Brazilian Afrofuturism, Climate Apocalypse, and Heuristic Function" explores the intersection of Brazilian Afrofuturism and climate apocalypse in the novels *(In)Verdades: Uma heroína negra mudará tudo* (2016) and *(R)Evolução: Eu e a verdade somos o ponto final* (2017) by Brazilian author and educator Lu Ain-Zaila. The novels present a politically and economically utopian space that responds to climate change and imagines a refoundation of Brazil. Brock contends that the rise of a Black middle class and affirmative action policies in Brazil have laid the groundwork for the boom of Afrofuturist literature and climate fiction as a response to the destructive discourses prioritizing profit over planetary well-being. The chapter examines the importance of imagination in counteracting dominant paradigms—for example, property rights, free markets, and free trade. It also explores how Ain-Zaila's novels advocate for climate and science fiction to challenge capital's monopolization of possible futures, collective action, resilience, economic innovation, and life after the apocalypse.

Lastly, two special contributions complete this volume: Jasmin Belmar Shagulian's "*Choike Pürrun*: The Mapuche People's Sacred Dance" and Igor Barreto's "The Paths Not Seen: How I Structured My Representation of Nature." These contributions function as ethnographic and poetic interventions intertwined with the overall purpose of *Contesting the Climate Unthinkable*: to unearth modes of cultural imagination that engage critically and creatively with our deranged time.

Belmar Shagulian's ethnographic intervention walks us through the significance of the *choike pürrun*, a ritualistic dance of the Mapuche people that represents the movements of the choike bird. The dance is part of the *nguillatún* ceremony, a ritual in which Mapuche people conjure fertility, validate the social order in their community, and seek synchronicity with Mother Earth. The dance symbolizes the complementary dualism in Mapuche cosmogony, with opposing forces of the feminine and masculine seeking balance. The effects of climate change and the politics of colonial and neoliberal oppression that have affected the Mapuche territories for centuries give even more relevance to Amerindian thought and cultural performance as an alternative to address the unthinkable. The piece emphasizes a holistic connection between humans and the cosmos, striving for kinship with all living entities to foreground the intangible presences in the territory ignored by the extractivist gaze.

Barreto's piece reflects on his poetic praxis. According to Barreto, his poems feature a force of implication with the peripheral and marginalized. The incorporation of narrative highlights, quotes, and paraphrases of lost texts into the poetic writing characterizes the lyrical atmosphere of Barreto's poetry. By doing so, the poems seek to contest the construction of decadent and neutral images of nature in contemporary poetry that lack engagement with the denseness of the world, including impure and contaminated evocations. In this intervention, Barreto shares poems from his book *El llano ciego* (The Blind Plain, translated by Rowena Hill), that call for representations of nature that connect the lyrical expression with our damaged planet. This contribution highlights the interstices in language and the environment and the attempt to bridge the representation of nature with a shared, ecologically sensitive experience.

Works Cited

Alimonda, Héctor. "La naturaleza colonizada. Una aproximación a la ecología política latinoamericana." *La naturaleza colonizada. Ecología política y minería en América Latina,* coordinated by Héctor Alimonda, CLACSO, 2011, pp. 21–58.

Andermann, Jens, Gabriel Giorgi, and Victoria Saramago. "Introduction." *Handbook of Latin American Environmental Aesthetics,* edited by Jens Andermann, Gabriel Giorgi and Victoria Saramago, De Gruyter, 2024, pp. 1–26.

Anderson, Mark, and Zélia Bora, editors. *Ecological Crisis and Cultural Representation in Latin America.* Lexington, 2018.

Barbas-Rhoden, Laura. *Ecological Imaginations in Latin American Fiction.* UP of Florida, 2011.

Bould, Mark. *The Anthropocene Unconscious: Climate Catastrophe Culture.* Verso, 2021.

Braidotti, Rosi. *Posthuman Knowledge.* Polity, 2019.

Bruce, Albert, and Davi Kopenawa. *The Falling Sky: Words of a Yanomami Shaman.* Harvard UP, 2013.

Castro, Azucena, and Gianfranco Selgas. "Sites of Situated Hope: Amazonian Rhythms, Unruly Caribbean Plants, and Post-Anthropocentric Gazes in Contemporary Latin American Cinema." *Journal of Latin American and Caribbean Studies,* vol. 31, no. 4, 2022, pp. 591–612.

de la Cadena, Marisol. *Earth Beings: Ecologies of Practice across Andean Worlds.* Duke UP, 2015.

de la Cadena, Marisol. "The Production of Other Knowledges and Its Tensions: From Andeanist Anthropology to Interculturalidad?" *Journal of World Anthropology Network* 1, 2005, pp. 13–33.

de la Cadena, Marisol, and Mario Blaser. *A World of Many Worlds.* Duke UP, 2018.

Delgado Ramos, Gian Carlo. "¿Por qué es importante la ecología política?" *Nueva Sociedad,* no. 244, 2013, pp. 47–60.

DeLoughrey, Elizabeth, Jill Didur, and Anthony Carrigan, editors. *Global Ecologies and the Environmental Humanities: Postcolonial Approaches.* Routledge, 2016.

Duchesne Winter, Juan R. *Plant Theory in Amazonian Literature.* Palgrave Macmillan, 2019.

Elias, Amy, and Christian Moraru, editors. *The Planetary Turn: Relationality and Geo-aesthetics in the Twenty-First Century.* Northwestern UP, 2015.

Escobar, Arturo. *Encountering Development: The Making and Unmaking of the Third World.* Princeton UP, 2011.

Escobar, Arturo. *La invención del desarrollo.* Universidad del Cauca, 2014.

Fornoff, Carolyn. *Subjunctive Aesthetics: Mexican Cultural Production in the Era of Climate Change.* Vanderbilt UP, 2024.

French, Jennifer, and Gisela Heffes. *The Latin American Ecocultural Reader.* Northwestern UP, 2020.

Ghosh, Amitav. *The Great Derangement: Climate Change and the Unthinkable.* U of Chicago P, 2016.

Giorgi, Gabriel. "Turning Territories into Temporalities: Three Aesthetic Methodologies." *GeoSemantics*, cluster edited by Azucena Castro and Estefanía Bournot, ASAP/J, 2023.
Gómez-Barris, Macarena. *The Extractive Zone: Social Ecologies and Decolonial Perspectives*. Duke UP, 2017.
Gudynas, Eduardo. "Buen Vivir: Today's Tomorrow." *Development*, vol. 54, no. 4, 2011, 441–47.
Haraway, Donna Jeanne. *Staying with the Trouble: Making Kin in the Chthulucene*. Duke UP, 2016.
Heffes, Gisela. *Políticas de la destrucción / Poéticas de la preservación. Apuntes para una lectura (eco)crítica del medio ambiente en América Latina*. Beatriz Viterbo Editora, 2013.
Kressner, Ilka, Ana María Mutis, and Elizabeth Pettinaroli. *Ecofictions, Ecorealities, and Slow Violence in Latin America and the Latinx World*. Routledge, 2019.
Leff, Enrique. "La Ecología Política en América Latina. Un campo en construcción." *Polis*, vol. 5, 2003, pp. 1–16.
Malm, Andreas. *The Progress of This Storm: Nature and Society in a Warming World*. Verso, 2018.
Merchant, Paul, and Lucy Bollington. *Latin American Culture and the Limits of the Human*. UP of Florida, 2020.
Merlinsky, Gabriela, and Paula Serafini. "Introducción." *Arte y Ecología Política*, edited by Gabriela Merlinsky and Paula Serafini, Universidad de Buenos Aires and CLACSO, 2020, pp. 11–26.
Moore, Jason W. *Capitalism in the Web of Life: Ecology and the Accumulation of Capital*. Verso, 2015.
Moore, Jason W. "There Is No Such Thing as Technological Accident: Cheap Nature, Climate Crisis, and Technological Impasse." *Technological Accidents, Accidental Technology*, edited by Joke Brouwer and Sjoerd van Tuinen, V2, 2023, pp. 1–21.
Morton, Timothy. *The Ecological Thought*. Harvard UP, 2010.
Mosse, Ramona. "Unthinkable Ecologies in Theatres of the Anthropocene." *Ecologies of Gender: Contemporary Nature Relations and the Nonhuman Turn*, edited by Susanne Lettow and Nessel Sabine, Routledge, 2022, pp. 103–16.
Nixon, Rob. *Slow Violence and the Environmentalism of the Poor*. Harvard UP, 2013.
Pratt, Mary Louise. *Planetary Longings*. Duke UP, 2022.
Rancière, Jacques. *The Politics of Aesthetics. The Distribution of the Sensible*. Continuum, 2004.
Rancière, Jacques. *The Politics of Literature*. Polity, 2011.
Rivera Garza, Cristina. "Geological Writings." *Latin American Literature in Transition 1980–2018*, edited by Mónica Szurmuk and Debra A. Castillo, Cambridge UP, 2022, pp. 15–29.
Robbins, Paul. *Political Ecology: A Critical Introduction*. Blackwell, 2004.
Saramago, Victoria. *Fictional Environments: Mimesis, Deforestation, and Development in Latin America*. Northwestern UP, 2021.

Spivak, Gayatri Chakravorty. *Death of a Discipline*. Columbia UP, 2003.
Svampa, Maristella. "Commodities Consensus: Neoextractivism and Enclosure of the Commons in Latin America." *South Atlantic Quarterly*, vol. 114, no. 1, January 2015, pp. 65–82. DOI: https://doi.org/10.1215/00382876-2831290.
Szeman, Imre, and Dominic Boyer. "Introduction." *Energy Humanities: An Anthology*, edited by Imre Szeman and Dominic Boyer, Johns Hopkins UP, 2017, pp. 1–13.

1

Reimagining the Land

Amerindian Aesthetics and Alternative Ecologies
for a Deranged World

1

Dying Well

Micropolitics of Life Against *Terricidio* in Mapuche Poetics

Montserrat Madariaga-Caro

In a 2019 interview, Moira Millán, Mapuche writer, *weichafe* (fighter), and, until recently, one of the spokespersons of the Movimiento de Mujeres y Diversidades Indígenas por el Buen Vivir, based in Argentina, was asked what she thought of climate change.[1] Moira Millán replied:[2] "Currently, the gaze is focused on a one-dimensional aspect of the problem. In the definition of climate change, there is an intentional reductionism.... It is not just about the relationship between production and consumption, but the anthropocentric model imposed by the dominant culture severed from the cosmic order.... The fight should not be against climate change, but against *terricidio*" ("La Lucha").[3]

For Millán and the members of the Movimiento, the climate catastrophes that are becoming more and more frequent are the result of the current global civilizing model imposed by ongoing forms of colonialism. This "predatory, colonial, racist, and patriarchal capitalist system" ("Documento" 238), as they describe it, in the day-to-day, is nothing less than a way of relating to the land that produces the violent death, slow or sudden, of human and more-than-human bodies. The Movimiento has described terricidio as "the murder of Mother Earth" ("Documento" 238)—specifically, "the murder of tangible ecosystems, of the spiritual ecosystem, of peoples, and all forms of life" ("Anexo 2" 241). Moreover, they propose that terricidio should be classified as a crime against humanity and nature ("Documento" 240). Thus, terricidio encompasses the current climate crisis but also refers to a much broader understanding of the environment because it centers on land from Indigenous perspectives, which do not separate humans from nature. As

Millán exposes, the fixation on climate change conceals the foundations of the crisis: white supremacy and neoliberal colonialism.[4] Acknowledging this white-settler foundation would require responding to Indigenous and racialized peoples' demands for justice and reparation.

In what follows, I argue that Mapuche poetics grow micropolitics of life that work as acts of justice in the face of terricidio when rooted in the AzMapu, the Mapuche land's justice ethos.[5] Particularly, I sustain that the video "Faumelisa Manquepillán—Poema La Materia—(Wetruwe Mapuche)" constitutes a multilayered poetic art piece that reclaims the Mapuche corpse's right to dying well.[6] The video holds micropolitics of life because it describes the decomposing of the corpse as a pleasurable act of renewing life; at the same time, it tributes to the Mapuche weichafe (fighters) that have been murdered defending the land. Overall, the video shows dying well as a form of poetic justice in defiance of terricidio.[7]

Faumelisa Manquepillán lives in the rural village of Puquiñe, near Valdivia city, in ancestral Williche Mapuche lands. Besides her two poetry books, *Sueño de mujer/Zomo Pewma* (2000) and *Lykan Küra Ñi Purrun. Danza de la piedra* (2017), Manquepillán is known for her work as a musician and *ülkantufe* (traditional singer), a basket weaver, and a sculptor. Wetruwe Mapuche is an independent project that promotes Mapuche poetry and music. It is run by Jaime Cuyanao, "Waikil," a well-known Mapuche rapper living in Santiago. The video capturing Manquepillán's poetic performance of "La materia" was recorded on March 21, 2013, by Jaime Cuyanao during International Poetry Day and uploaded to YouTube in December of the same year. In the following sections, I first delve into the concept of the "micropolitics of life" and the Mapuche AzMapu while also providing context for terricidio within the framework of Chile as a settler nation. Then, I analyze Manquepillán's poem and performance, focusing on the intricate relations between the corpse and the land. Lastly, I look into Wetruwe Mapuche's opening sequence to the video, situating the poetic performance within the broader context of the Mapuche defense of the land and death by terricidio.

Mapuche poetics has taught me to be literally and imaginatively attentive to land—its rooted and submerged ethics, sedimented justice, and subterranean wisdom of bodies turning into humus. I aim not to disclose the Mapuche approaches to death to gain knowledge of their cultural practices. Instead, through a reading of the audiovisual art piece, I aim to illuminate how neoliberal colonialism continues to dehumanize the Mapuche even in death. Moreover, the intention is to illustrate how poetic actions can uphold the dignity of the departed.

Micropolitics of Life and the AzMapu in the Settler Nation

I argue that Mapuche poetics enact diverse micropolitics of life—that is, actions based on emotional and sensorial knowledge gained through multiple quotidian interactions with a specific land's diverse human and more-than-human bodies. Their purpose is to (auto)regenerate, strengthen, and recuperate life from the settler-colonial, capitalistic, patriarchal, and racist order that provokes terricidio. Thus, poetics that hold micropolitics of life nourish the conditions and relationships that make life possible in specific lands.

My elaboration of "micropolitics of life" as a concept and an analytical tool applied to poetics grows primarily from the writings of Héctor Nahuelpán (Mapuche) and Suely Rolnik (Brazilian). Nahuelpán defines "Mapuche micropolitics of resistance" as actions that strategically contest power relations and undermine conditions of racial and historical oppression in everyday life (95). These practices are rooted in personal and collective self-governance, territorial control, and a profound relation with the Mapu (116–18). Rolnik views micropolitical insurgency as aiming to decolonize the unconscious mind and reclaim life and its creative energy from the oppressive "colonial-capitalist" order (112–20). "Artistic operations" help heal from trauma, generating the language and actions necessary for envisioning and experiencing new worlds (119).[8] Nahuelpán's autoethnographic perspective and Rolnik's psychological insights have resonated in my approach to Mapuche poetics as acts that invigorate land and life. Focused on the Willi Mapu, the south of Gnulu Mapu, the ancestral Mapuche land to the west of the Andes mountains, I have found that the works of Cristian Antillanca, Leonel Lienlaf, Faumelisa Manquepillán, Roxana Miranda Rupailaf, Adriana Paredes Pinda, and Kütral Vargas Huaiquimilla are infused by a body/land knowledge that, through the aesthetic emotions they evoke, open pathways, not of worlds to come, but of a more liberated today. Mapuche poetics reflect a nonlinear space/time that is not utopian because it coexists with the terricidal order. Moreover, these Mapuche poetics go beyond settler-colonial justice beliefs and institutions as they follow the AzMapu, the Mapuche land's justice ethos.

The notion of "AzMapu" comes from the knowledge of the Mapuche elders, the *kimche,* and it cannot be equated to a settler concept of justice; however, in the context of colonialism, its validity is grounded in processes of resistance and self-determination (Melin et al., *AzMapu* 22). Thus, the AzMapu can be understood as the "Mapuche inherent rights" (14) from a Western framework, and expansively as how the Mapu organizes life. *Mapu* is a polysemic word that can refer to the universe or a specific territory, encompassing multiple material and spiritual spaces within (Comunidad 16). Each *che*, person, family, and community are formed by the AzMapu of their place of origin (Chihuailaf 48–49). According to Mapuche anthropologist Jimena Pichinao, the AzMapu corresponds to the "way of being of a territory" (101).[9] The AzMapu regulates responsibilities to place and social relations, including plants, nonhuman animals, spiritual forces, and ancestors (Melin et al., *AzMapu* 21, 26, and 27). This ethos informs Mapuche justice because restitution represents a fundamental and constitutive element of the AzMapu; when someone fails to live by it, the land sanctions them and asks for reparations (Melin et al., *AzMapu* 21). This enactment of the AzMapu resonates with what activist Dina Gilio-Whitaker (Colville Confederate Tribes) affirms: "In essence, justice for Indigenous peoples is about restoring balance in relationships that are out of balance" (26). Therefore, poetic micropolitical acts that strengthen life amid terricidio follow the AzMapu.

Manquepillán's and Wetruwe Mapuche's art piece honors the AzMapu by representing dying well as a life-giving act and reclaiming the Mapuche right to it. Dying well can be understood as a continuum of living well or *küme mongen* in Mapuzungun.[10] In conversation with Manquepillán, she conveyed that living well entails living in harmony with the *itrofill mongen*, which encompasses all forms of life, including spiritual entities, across the various dimensions of the Mapu (Melin et al., *Cartografía* 47–48), sometimes inaccurately translated as biodiversity. For Manquepillán, the meaning of *küme mongen* can be found in Mapuche poetry and Mapuche hope for the end of colonial violence. The perpetrators of terricidio, mainly but not only nation-states and corporations, violate the AzMapu, damaging the social dynamics among humans and more-than-humans and, thus, obstructing the *küme mongen*. When settlers occupy, privatize, and fence the land for factories, churches, and large estates, they weaken the spiritual and physical balance among all bodies of life in a territory.

As a nation-state built by a criollo society (the direct descendants of Spanish colonizers), Chile is a settler state founded on the principle of private property to exploit land and human and more-than-human bodies. In contrast, Pichinao asserts that every territory of the Mapu has an "owner-guard," a *ghe*, a spiritual entity, for lack of better words, that protects the land. The Mapuche can "co-own" the land with the ghe, in the sense of reciprocal care, following the AzMapu (99). In fulfilling the role of protecting the Mapu, the Mapuche halted Spaniard colonization and maintained political, economic, and territorial autonomy south of the Biobío River. During the first half of the nineteenth century, Chile's nascent republic continued the tradition of diplomatic relations through parliaments started by Mapuche and Spaniards in 1641. However, escalating tensions regarding Chile's territorial sovereignty and citizenship culminated in a military invasion from 1860 to 1883 (Mariman 84–86). Historian Ana Vásquez contends that in the first half of the twentieth century, Chilean law aimed to undermine Mapuche's self-governance and everyday internal practices (143–44), which affected their land tenure. In other words, it aimed to break Mapuche's understanding of ownership as caring for the land.

During Augusto Pinochet's dictatorship (1973–90), new laws facilitated further colonization of Mapuche territory, especially for developing farming, forestry, and hydroelectric industries (Rodenkirchen 253–60).[11] Since the return of democracy in 1990, the Chilean state has responded to Mapuche demands for justice within the settler-colonial mindset of neoliberal multiculturalism.[12] While Chile's recognition policy celebrates and commodifies Mapuche culture, the courts of law criminalize Mapuche's actions to recuperate ancestral lands (Del Valle Rojas 223). State violence in rural Mapuche communities is now commonplace, characterized by disproportionate force and discrimination, leading to evictions, unlawful beatings, violent arrests, and shootings at homes, among other abuses (Luneke et al. 3). Twenty-five Mapuche have been killed since the 1990s, with twelve deaths directly attributed to the Chilean armed forces. Others have been fatally run over or shot by private guards and gunmen (Delgado). Almost all of them are known to have participated in land reclamations.[13] Since 2021, state agents and hegemonic media have been instilling in public opinion the fake existence of "narco-terrorism networks in emblematic communities of the Mapuche resistance" (Nahuelpán et al.). Last year, the Chilean congress approved Law 21633, commonly known as the "usurpation law," penalizing illegal occupation of public and private properties, despite Mapuche congress members voicing its potential use to criminalize Mapuche land reclamations (Lee). Ultimately, Chile's executive, legislative, judicial, and "fourth estate" defend the settler-colonial structure of the nation and an economy based on terricidio.

Manquepillán's performance at the poetic gala in Santiago brings the Mapu to the core of the settler state. The event was held in Providencia (Teperman), a commune built by criollo families, whose inherited wealth was obtained based on stolen Indigenous land and terricidio. The gala took place at the town hall, also known as the Falabella Palace, an icon of criollo prosperity during the first half of the twentieth century—a testament to colonial violence. As I will illustrate, the audiovisual poetic piece embodies the Mapu, reclaiming the sovereignty of the corpse and the land over the artifice of settler's laws. The video reproduces images of the AzMapu that work as micropolitics of life against terricidio as it embraces the poetic justice of dying well. What follows is a multilayered analysis of the video as a complex poetic artifact: I interpret the poem's meanings, Manquepillán's performance, and Wetruwe Mapuche's framing of the poetic act.

Death as Nurturing the Mapu's Corporeality

Manquepillán's poem "La materia" is a homage to the act of dying well as a vital dimension of the küme mongen. The verses of the poem speak of a pleasurable psychosomatic experience of death: "Dejo que mi cuerpo frío se sumerja entre los maquis del camino, / que el olor putrefacto de mis vísceras sea magneto para aves de carroña / y me inviten a volar en sus entrañas / Dejo que mi boca y mi nariz sean alimento y cuna de moscas y sus larvas / Dejo que mis fluidos sean agua pura entre las nubes / Dejo que mis huesos sigan tomando el sol por las mañanas / y que por las noches mi espíritu sonría y dance al compás de la vida / con mi calavera" [I allow my cold body to submerge among the maquis of the road, / may the rotten smell of my entrails be a magnet for carrion birds / and invite me to fly into their gut / I allow my mouth and my nose to be food and a cradle for flies and their larvae / I allow my fluids to be pure water between the clouds / I allow my bones to continue to sunbathe in the mornings / and at night my spirit smiles and dances to the beat of life / with my skull] (Manquepillán and Wetruwe Mapuche 1:57–2:38).

Far from transmitting a macabre tone, there is joy in the actions of these verses: to dive, to fly, to be pure water, to sunbathe, to smile, and to dance. The repetition of the verb *dejar*, to allow, shows that the speaker tenderly and willingly surrenders to the body's changes. The verses focus on the physical transformations of the corpse, creating sensory images that capture the purpose of the AzMapu by depicting harmonious (necrological) interactions among the diverse lives of the land.

As raw material for life, the poem presents death as a generous act of giving that invigorates the land. The decomposing corpse submerged among the maquis, a Mapuche native plant, breaks down into humus, fertile organic matter. Thus, the cadaver becomes nutrients for the *lawen*, the medicinal plants of the Mapu that cure sick bodies, completing a virtuous circle. Likewise, the cadaver's entrails, mouth, and nose are food for birds and flies, as if the microbiological agents of the decomposing body were mothering and nurturing the Mapu's multiple offspring. These poetic images align with what Pichinao affirms, that humans, nonhumans, and spiritual forces inhabiting a place are called to practice a co-caring of the Mapu to guarantee the küme mogen (98–99). In this sense, dying well, as an event that invigorates the land, can be read as micropolitics of life against terricidio because it heals the soil from the depletion and harm caused by industrial agriculture, large-scale forestry, and other settler businesses.

The poem recognizes the corpse's shared materiality with the Mapu. Furthermore, it recognizes the Mapu as matter, the source of all that naturally exists. "Everything is *Mapu;* everything is matter," says kimche Juan Ñanculef (51). The human body is matter that returns to the land once deceased, and the spirit (*püllü*) is also matter, "the subtlety of the land" (51) that transcends to another (material) dimension of the Mapu, the Mallew-Mapu (62). Through its alliteration and cadence, the poem suggests that the lyrical I, as a representation of the spirit, expresses matter itself and voices the Mapu.

Manquepillán's worldview in "La materia" agrees with notions of new materialism in that the poem decenters anthropocentric binaries between human and nature, life and matter, life and death. However, for Manquepillán, poetry is the very stuff, the matter, that organisms are made of: "I think that the stone, the fiber, the wood, are also poetry, and there is no difference" (Manquepillán et al. 106). In the poem, the present tense verbs materialize the mortuary desires; the poetic voice gives birth to the transmutation of matter. The patterning of the rhythm and the soothing repeating "s" sounds transmit the wellness of the body's decay. Poetry shapes the matter, shapes the Mapu, in Manquepillán's work. Furthermore, poetry serves as the breath and heartbeat of the land, as her performance demonstrates.

The performance creates an acoustic corporeality of the Mapuche land. Manquepillán begins her presentation by chanting an *ül* (song) and playing a *kultrun*, a percussion instrument used in Mapuche ceremonies. The kultrun represents the Mapuche universe or the Meli Witran Mapu, the four sides of the Earth that orient Mapuche territorial identities (Chuhuailaf 36–37). In the performance, the kultrun's reverberating sounds resemble the beating of a heart, reproducing the rhythm of the Mapu's body. In tandem with this cadence, Manquepillán sings in Mapuzungun, letting the sounds of the Mapu fill the waving air with its messages. Singing is a tradition in Manquepillán's family used to pass on knowledge: "I remember my grandparents singing in *Mapuzungun* many times, narrating their path, their life trajectory in their singing" (Manquepillán et al. 106). In her recital, through the beating of the kultrun and the breath of the ül, Manquepillán embodies the land. She creates an acoustic ecology of togetherness, paying tribute to the Mapu and those who have violently died. For most of the audience, these codes are unintelligible because, as part of the genocidal practices of the settler state, the Mapuche language and musical instruments were forbidden and excluded from public life during almost all of the twentieth century. The cognitive border raised by Manquepillán's performance demonstrates the persistence of an unassimilated AzMapu.[14]

The recital poetically renews the Mapu's life cycle in a public ritual at the heart of the settler-colonial nation. It also advocates for dying well among the Mapuche, invoking the AzMapu. Ultimately, Manquepillán reclaims Mapuche corpses from death by terricidio. The poems' images of a slow, generous, and peaceful letting go of the human canvas and going back to the "original matter" contrasts with the experience of many Mapuche land defenders murdered due to white-settler entitlement to Indigenous life and death.

Reclaiming the Dead from Terricidio

In light of Wetruwe Mapuche's framing of Manquepillán's poetic performance, I interpret the whole art piece as a vindication ritual for Mapuche's unjust deaths. The audiovisual media of Wetruwe Mapuche places Manquepillán's performance in the symbolic sphere of the Mapuche politics of self-determination and resistance. The video starts with an opening sequence of a Mapuche person swinging a *wetruwe*. In the Mapuche language, *wetruwe* means *boleadora*, a throwing weapon consisting of a rope with weights attached to it or a rope used as a sling. Traditionally, it is considered a tool to hunt animals. However, in the context of colonialism, the Mapuche have used it against invaders, more recently, the police and other perpetrators of terricidio. In their blog, Wetruwe Mapuche states they want to "use culture, music, and poetry as a *wetruwe*." They say these are weapons that "recover our *Kimün* (culture), *Rakizuam* (thought), and *Mapuzungun*" (Wetruwe). This statement signifies Manquepillán's performance as an act of recovery.

The introduction of the video evokes the figure of the *weichafe*, the Mapuche fighter who defends the land from new settlers, big landowners, and hydroelectric and timber corporations that contribute to the Mapu's terricidio.[15] The footage shows a short-haired young person wearing traditionally masculine clothing like the *trarilongko* (headband) and the *maküñ* (cloak). The images are accompanied by the sound of a *trutruka*, a wind instrument currently used by the Mapuche mainly in the *nguillatún* (prayer ceremony) and the *purrun* (ceremonial dances) ("Trutrukero" 3:03–3:20). Yet, due to colonial chronicles that have referenced the trutruka as part of the Mapuche war repertoire, this instrument is associated with Mapuche resistance today (Soto-Silva et al. 10–12). The images of the young weichafe with a wetruwe together with the soundwaves of the trutruka invoke the physical and spiritual Mapuche *newen:* the strength to attack, resist, and defend the Mapu as the AzMapu orders. Thus, in the video, the weichafe newen, the fighter's strength, is transferred to the poetic performance, imbuing it with a tribute to the murdered Mapuche who lived by the AzMapu.

Death by terricidio, in this context, is a crime motivated by settlers' encroachment upon Mapuche land, aiming to exploit it for profit by eliminating the people who refuse to give it up. About the constant murdering of Mapuche weichafe, Manquepillán said in an interview: "This form of derision and dispossession that the state does with our people is excruciating" (Manquepillán et al. 108). At the time of the interview, in 2018, the Mapuche were mourning the death of twenty-four-year-old Camilo Catrillanca, shot in the back by the police agent Carlos Alarcón in the rural Mapuche community Temucuicui. Catrillanca was the grandson of a *lonko,* a Mapuche political authority, and had publicly defended his community's rights to the land since high school. The police had been surveilling him for a year before his murder (Sepúlveda). His death is a consequence of the militarization of the Nagche Mapuche territories and the "Jungle Command," a special police force trained in Colombia to fight terrorism (Nahuelcheo et al. 59). A large contingent of this group was searching for car theft suspects when they encountered Catrillanca driving a tractor accompanied by a fifteen-year-old. The policemen involved in the crime first declared that they were attacked and claimed that Catrillanca had died after crossing into the line of fire (Basadre et al.). Later, police videos, initially concealed, revealed the facts. Catrillanca's assassination exemplifies the cruel contempt toward Mapuche life by the settler-state agents who commit terricidio.[16]

As with the land, the settler state exercises its racist sovereignty over the Mapuche corpse. When a person is killed, the police take the dead body as part of their investigation procedure. In the context of death by terricidio, the retention of the corpse constitutes another layer of violence because the body is, in a way, kidnapped and held hostage by the state. The Mapuche cadaver is subdued to the settler-colonial laws, scientific scrutiny, and administrative protocols, becoming evidence, a lifeless thing in the hands of the settler's agents. This capture of the deceased disrupts and delays the process by which the corpse gives life to the land and the body and spirit transform their materiality, the process Manquepillán sings to. It also delays the deceased's farewell and the ceremony that honors this passage.

The Indigenous cadaver has been a territory of exploration and exploitation, especially since the development of scientism in France during the nineteenth century (Huinca 91). As a result, Indigenous bodies and body parts, without consent, are retained in Western spaces such as laboratories, museums, world exhibitions, universities, and private collections. These spaces inscribe them into settler narratives and poetics that make and remake the settler state. For instance, in 1882, the National Museum of Chile received as a gift an "Indian skeleton and a skull" (Alegría et al. 9), both found together near Osorno city, in ancestral Mapuche lands. Mapuche researcher Herson Huinca asserts that human sciences served to legitimize colonization by demonstrating with prejudicial measurements the supposed European racial superiority (93–94). To this day, hegemonic science promotes the idea of the inevitable extinction of Indigenous peoples to justify using their bodies, alive or dead, as resources for scientific experimentation (TallBear 182–85). In Chile, human remains and objects exhumed from Indigenous cemeteries are regarded as part of the nation's archaeological heritage, which means that the state and private institutions appoint themselves the custodians of any native ancestors and sacred objects found. In this sense, heritage constitutes another form of Indigenous dispossession.

The scientific, administrative, and police violence over the dead Mapuche body are not explicit in Manquepillán's performance of the poem "La materia;" still, it constitutes the conditions under which she lives and creates. Mapuche Wetruwe's intro of the video reminds the audience of the constant defense of the land and its costs. When understanding this context, the sense of freedom in the poem intensifies and shows its refusal to terricidio. The poetic performance liberates the Mapuche corpse from the disaffected space of the morgue. It holds the Mapuche micropolitics of life because, as humus in formation, a rebellious love of Mapuche life and land breaks down violence into nutritious tenderness.

Conclusions

Throughout these pages, I analyzed the micropolitics of life against terricidio in the video "Faumelisa Manquepillán—Poema La Materia—(Wetruwe Mapuche)." This multilayered work is rooted in the justice ethos of the collective lived experience of the AzMapu. Furthermore, the video conveys Mapuche poetics that honor dying well as an integral part of living well, reflecting a caring relationship with all life forms of the Mapu, as pursued by the AzMapu.

Attending to the intrinsic relation between the material, the spiritual, and the poetic realms of life, as well as the need to protect life, the poetic piece pays tribute to the tenacious renovation of the Mapuche Mapu. The video is a homage to a joyfully regenerated life that continues to resist settler colonialism from the depths of its microbiology to the direct actions of the weichafe. The piece enacts micropolitics of life by strengthening the lands' breath with a poetic justice ritual that mourns the dead and forges the foundations for new Mapuche life.

The concept of terricidio illuminates the commonly unspoken causes of the current climate crisis: neoliberal colonialism and white supremacy. Terricidio corresponds to the murdering of the land, its ecosystems, its spiritual forces, and the Indigenous, racialized, and feminized people who care for it by the wealthiest on planet Earth. By acknowledging terricidio, we can identify the poetics restoring life where there is violent death.

Notes

1 My thanks to Dr. Luis Cárcamo-Huechante for introducing me to the concept of *terricidio,* Dr. Ruth Rubio-Gilbertson for revisions to the first draft, the anonymous peer reviewers, and editors of this volume.
2 In English, Movement of Indigenous Women and Nonbinary Folks for Living Well.
3 All translations are mine.
4 I follow Shannon Speed, who affirms that nation-states of Latin America practice neoliberal colonialism, which "continues the work of forcing Indigenous peoples to slave on their own expropriated lands for others' benefit while deploying interlocking race and gender tropes against them in ways that further suppress their inherent sovereignty as peoples and render their dispossession acceptable" (788).
5 "Mapuche" means people of the Mapu, the land, and it is not pluralized as "Mapuches."
6 The video can be played at https://www.youtube.com/watch?v=m6dB2yqQtjU.
7 A seed of some of the ideas presented here can be found in my article "'Te llaman en lenguas raulíes y alerzarias': poéticas de mujeres por el cuidado de la mapu en territorio Mapuche-Williche," published in the book *Hacer cantar la maravilla: plantas medicinales en poemas de mujeres Chile-Wallmapu XX-XXI* (Fondo de Cultura Económica, 2022) edited by Rubí Carreño, Claudia Rodríguez, and Estela Imigo. In it, I study poems by Adriana Paredes Pinda and Faumelisa Manquepillán to argue that the complicity of women and plants acts as medicine that resists settler colonialism.

8   I'm also dialoguing with Dian Million's "felt theory" (53); Silvia Rivera Cusicanqui's "politics of subsistence" (*Un mundo* 142); Leanne Simpson's "Nishnaabeg radical resurgence" (31); and James Scott's "infrapolitics." While "infrapolitics" underlines everyday life as the "unobtrusive realm of political struggle" (183) in reference to the subaltern's "disguised, low-profile, undeclared resistance" (298), the term "micropolitics" centers on the relationships between human people and the land and its intention of recuperating life and bonds between bodies of life of particular places.

9   The concept *AzMapu* can also be written as *az mapu* and *Ad Mapu*. I follow the orthography in the book *AzMapu. Una aproximación al Sistema Normativo Mapuche desde el Rakizuam y el Derecho Propio* (2016).

10 Over approximately the past fifteen years, the term *küme mongen* has expanded its meaning as the paradigm of *buen vivir* ("good living," "living well") has gained traction in Latin America. *Buen vivir* revolves around practices of harmonious relationship between humans and nature. The term derived from the Kichwa concept of *sumak kawsay*, popularized by Ecuador's Indigenous movement of the 1990s to emphasize self-determination within specific territories (Altmann 9). Buen vivir now finds application among community-based feminists, rural and Indigenous-led groups like the Movimiento de Mujeres y Diversidades Indígenas por el Buen Vivir. It also has been academically theorized as a critique of and an alternative to the Western capitalist development ideology (Gudynas 2–3). In 2008, Ecuador incorporated buen vivir / sumak kawsay into its constitution, followed by Bolivia a year later, drawing from the Aymara concept of *suma qamaña* (3–5). However, its adoption in academic and state contexts has faced criticism. Philipp Altmann contends that buen vivir was co-opted and distorted from Ecuador's Indigenous peoples as part of a colonizing epistemic movement (1–2). Furthermore, Ecuador and Bolivia have been criticized for failing to uphold the buen vivir by supporting extractivist projects against Indigenous communities' will. My reference to the küme mongen does not draw from the academic and state-sanctioned versions of buen vivir; instead, it aligns with its meanings anchored in self-determination as shown in the analysis of the poetic piece.

11 In 1979, Pinochet's decree-law 2568 facilitated Indigenous land theft by recognizing anyone residing on communal lands as Indigenous, permitting the division of these lands upon request. This enabled local settlers posing as *comuneros* to claim parcels once communal lands were divided, stripping them of Indigenous status (Rodenkirchen 253–54). Other regulations since 1974, including the Forestry Law (decree-law 701), generously subsidize private companies to exploit nonnative forest species like pine and eucalyptus.

12 Charles Hale understands "neoliberal multiculturalism" in the late twentieth and early twenty-first centuries in Latin America "as a regime of governance produced in part by effective mobilizations from below for rights grounded in cultural difference, and in part from preemptive moves by dominant actors and institutions to limit and manage these rights, and the people who would bear them" (619). Rivera Cusicanqui affirms that across Latin America the "technocrats" implemented

"an ornamental and symbolic multiculturalism with prescriptions such as 'ethno-tourism' and 'eco-tourism'" ("*Ch'ixinakax*" 98) that concealed new forms of colonization. In sum, nation-states capitalize on Indigenous cultures while maintaining and even exacerbating structural racism and colonial inequalities (Richards 65–66; Zapata 18–28).

13 "Land reclamations" or "territorial recuperations" denote instances where Mapuche communities assert autonomy from the State, establishing settlements without authorization on their ancestral lands, presently owned by affluent criollo settlers, timber, and hydroelectric corporations (Llaitul 357). Originating in 1971 during Salvador Allende's agrarian reform, these actions persisted through the dictatorship era. With the restoration of democracy (1990), the autonomist organization CAM (Coordinadora Arauco-Malleco) advocates for land recuperations and direct action to thwart capitalist endeavors, aiming for Mapuche national liberation. Various Mapuche communities across the historical Mapuche territory have responded to this call (Pairacan and Urrutia).

14 Following xwélmexw scholar and artist Dylan Robinson's ideas of "settler colonial listening" (3), I do not include a translation of the *ül*. Robinson theorizes on the "hungry listening" of settlers who want to have access to all Indigenous creations as a form of "Indigenous knowledge extraction" (23). If Manquepillán wanted the audience to understand the words, she would have sung the ül in Spanish as well, as she does when that is her will.

15 The weichafe is not necessarily a "man"; however, due to the patriarchal gender/sexual binary more males are recognized as such than females.

16 In 2021, Alarcón was sentenced to sixteen years in prison, marking the first instance of a policeman being incarcerated for murdering a Mapuche. Catrillanca's family wanted a life sentence.

Works Cited

Alegría, Luis, et al. "Momias, cráneos y caníbales. Lo indígena en las políticas de 'exhibición' del Estado chileno a fines del siglo XIX." *Nuevo Mundo Mundos Nuevos, Débats*, 2009. Open Edition Journals, doi.org/10.4000/nuevomundo.53063.

Altmann, Philipp. "The Commons as Colonisation: The Well-Intentioned Appropriation of Buen Vivir." *Bulletin of Latin American Research*, 2019, https://doi.org/10.1111/blar.12941.

Basadre, Pablo, and Equipo CIPER. "Muerte de Catrillanca: así se inventó la versión falsa de Carabineros." Ciperchile.cl, 1 Feb. 2019, https://www.ciperchile.cl/2019/02/01/muerte-de-catrillanca-asi-se-invento-la-version-falsa-de-carabineros/

Chihuailaf, Elicura. *Recado confidencial a los chilenos*. 2nd ed., LOM, 2015.

Comunidad de Historia Mapuche. "Awükan ka kixankan zugu. Kiñeke rakizuam." *Awukan ka kaxankan zugu wajmapu mew. Violencias coloniales en Wajmapu*, edited by Enrique Antileo et al., Ediciones Comunidad de Historia Mapuche, 2015, p. 16.

Del Valle Rojas, Carlos. "La crisis de la interculturalidad en la administración de la justicia en los tribunales del sur de Chile y el rol de la producción del enemigo íntimo-interno en la industria cultural." *Justicia e interculturalidad. Análisis y pensamiento plural en América y Europa,* coordinated by Marianella Ledesma, Centro de Estudios Constitucionales, Tribunal Constitucional del Perú, 2017, pp. 221–50.

Delgado, Camila. "Mapuche asesinados en democracia: cuando el manto de impunidad recorre a todos los gobiernos en Chile." *La Izquierda Diario,* 3 Jan. 2022, https://www.laizquierdadiario.cl/Mapuche-asesinados-en-democracia-Cuando-el-manto-de-impunidad-recorre-a-todos-los-gobiernos-en. Accessed 10 Dec. 2022.

Gilio-Whitaker, Dina. *As Long as Grass Grows: The Indigenous Fight for Environmental Justice, From Colonization to Standing Rock.* Beacon, 2019.

Gudynas, Eduardo. "Buen vivir: generando alternativas al desarrollo." *America Latina en Movimiento,* no. 462, 2011, pp. 1–20.

Hale, Charles. "Using and Refusing the law: Indigenous Struggles and Legal Strategies After Neoliberal Multiculturalism." *American Anthropologist,* vol. 122, no. 3, 2020, pp. 618–31.

Huinca, Herson. "Los Mapuche del Jardín de Aclimatación de París en 1883: objetos de la ciencia colonial y políticas de investigación contemporáneas." *Ta iñ fijke xipa rakizuameluwün. Historia, colonialismo y resistencia desde el país Mapuche,* edited by Héctor Nahuelpán et al., Santiago, Ediciones Comunidad de Historia Mapuche, 2013, pp. 89–118.

Lee, Lun. "'Afecta al reclamo de tierras ancestrales': las consecuencias de la Ley de Usurpaciones que advierten desde el mundo mapuche." Interferencia.cl, 10 July 2023, https://interferencia.cl/articulos/afecta-al-reclamo-de-tierras-ancestrales-las-consecuencias-de-la-ley-de-usurpaciones-que

Llaitul, Héctor. "¡El territorio no se compra, se recupera . . . ! ¿Compra-venta, expropiación control territorial?" *Conflictos étnicos, sociales y económicos. Araucanía 1900–2014,* edited by Jorge Pinto, Pehuén, 2015, pp. 353–58.

Luneke, Alejandra, et al. "Policía militarizada en Chile: claves para comprender la violencia policial estatal en la relación al conflicto mapuche," *Anuario del Conflicto Social,* no. 13, 2022, p. e-40763, https://doi.org/10.1344/ACS2022.13.8.

Manquepillán, Faumelisa. Interview. Conducted by Montserrat Madariaga-Caro. 23 March 2024.

Manquepillán, Faumelisa, Cecilia González, and Rocío Viñas. "La Gempin. Entrevista a Faumelisa Manquepillán." *Exlibris,* no. 8, 2019, pp. 104–9.

Manquepillán, Faumelisa, and Wetruwe Mapuche. "Faumelisa Manquepillán—Poema La Materia—(Wetruwe Mapuche)." YouTube, uploaded by Wetruwe Mapuche, 17 Dec. 2013, https://www.youtube.com/watch?v=m6dB2yqQtjU.

Mariman, Pablo. "La República y los Mapuche: 1819–1828." *Ta iñ fijke xipa rakizuameluwün. Historia, colonialismo y resistencia desde el país Mapuche,* edited by Héctor Nahuelpán et al., Ediciones Comunidad de Historia Mapuche, pp. 63–87.

Melin, Miguel, et al. *AzMapu. Una aproximación al sistema normativo Mapuche desde el rakizuam y el derecho propio*. Instituto Nacional de Derechos Humanos and Unión Europea, 2016.

Melin, Miguel, Pablo Mansilla Quiñones, and Manuela Royo. *Cartografía cultural del wallmapu: Elementos para descolonizar el mapa en territorio mapuche*. LOM Ediciones, 2019.

Millán, Moira. "La lucha no debe ser contra el 'cambio climático' sino contra el terricidio." Entrevista por Lucía Cholakián Herrera, Nodal, https://www.nodal.am/2019/10/moira-millan-lideresa-mapuche-la-lucha-no-debe-ser-contra-el-cambio-climatico-sino-contra-el-terricidio/. Accessed 10 May 2021.

Million, Dian. "Felt Theory: An Indigenous Feminist Approach to Affect and History." *Wicazo Sa Review*, vol. 24, no. 2, 2009, pp. 53–76.

Movimiento de Mujeres y Diversidades Indígenas por el Buen Vivir. "Anexo 2." *Repensar el Sur. Las luchas del pueblo Mapuche*, coordinated by Raúl Zibechi and Edgars Martínez, Clacso and Retos Cooperativa Editorial, 2020, pp. 241–42.

Movimiento de Mujeres y Diversidades Indígenas por el Buen Vivir. "Documento del Campamento Climático." *Repensar el Sur. Las luchas del pueblo Mapuche*, coordinated by Raúl Zibechi and Edgars Martínez, Clacso and Retos Cooperativa Editorial, 2020, pp. 237–40.

Nahuelpán, Héctor. "Micropolíticas mapuche contra el despojo en el Chile neoliberal. La disputa por el *lafkenmapu* (territorio costero) en Mehuín." *Izquierdas*, vol. 30, 2016, pp. 89–123.

Nahuelpán, Héctor, et al. "¿Para qué se construyó la idea del narcoterrorismo en Wallmapu?" Ciper, 14 July 2021, https://www.ciperchile.cl/2021/07/14/para-que-se-construyo-la-idea-del-narcoterrorismo-en-wallmapu/. Accessed 30 Sep. 2021.

Nahuelcheo, Pamela, et al. "Crímenes y montaje como política indígena." *Anuario Del Conflicto Social*, no. 9, 2020, https://doi.org/10.1344/ACS2019.9.4.

Ñanculef, Juan. *Tayiñ Mapuche Kimün Epistemología Mapuche. Sabiduría y Conocimientos*. Universidad de Chile, 2016.

Pairacan, Fernando, and Marie Juliette Urrutia. "La rebelión permanente: una interpretación de levantamientos mapuche bajo el colonialismo chileno." *Radical Americas*, vol. 6, no. 1, 2021, https://doi.org/10.14324/111.444.ra.2021.v6.1.012.es.

Pichinao, Jimena. "La mercantilización del Mapuche Mapu (tierras mapuche). Hacia la expoliación absoluta." *Awukan ka kaxankan zugu wajmapu mew. Violencias coloniales en Wajmapu*, edited by Enrique Antileo et al., Santiago, Ediciones Comunidad de Historia Mapuche, 2015, pp. 87–106.

Richards, Patricia. "Of Indians and Terrorists: How the State and Local Elites Construct the Mapuche in Neoliberal Multicultural Chile." *Journal of Latin American Studies*, vol. 42, no.1, 2010, pp. 59–90.

Rivera Cusicanqui, Silvia. "*Ch'ixinakax utxiwa*: A Reflection on the Practices and Discourses of Decolonizatio." *South Atlantic Quarterly*, vol. 111, no. 1, 2012, pp. 95–109.

Rivera Cusicanqui, Silvia. *Un mundo ch'ixi es posible. Ensayos desde un presente en crisis*. Tinta Limón, 2019.

Robinson, Dylan. *Hungry Listening: Resonant Theory for Indigenous Sound Studies.* U of Minnesota P, 2020.

Rodenkirchen, Alina. "Memorias Mapuche en la continuidad colonial: experiencias durante la dictadura militar chilena (1973–1990)." *Awukan ka kaxankan zugu wajmapu mew. Violencias coloniales en Wajmapu,* edited by Enrique Antileo et al., Ediciones Comunidad de Historia Mapuche, 2015, pp. 239–70.

Rolnik, Suely. *Esferas de la insurrección. Apuntes para descolonizar el inconsciente.* Tinta Limón, 2019.

Scott, James. *Domination and the Art of Resistance: Hidden Transcripts.* Yale UP, 1990.

Sepúlveda, Nicolás. "Informe policial secreto: Camilo Catrillanca estaba en la mira de Carabineros." Ciperchile.cl, 27 Nov. 2018, https://www.ciperchile.cl/2018/11/27/informe-policial-secreto-camilo-catrillanca-estaba-en-la-mira-de-carabineros/.

Simpson, Leanne Betasamosake. *As We Have Always Done: Indigenous Freedom Through Radical Resistance.* U of Minnesota P, 2017.

Soto-Silva, Ignacio, et al. "The Trutruka Playing as a Representation of Mapuche Resistance in the Urban Popular Music Scene in the Region of Los Lagos, Chile." *Per Musi,* no. 42, 2022, pp.1–25, https://doi.org/10.35699/2317-6377.2022.39490.

Speed, Shannon. "Structures of Settler Capitalism in Abya Yala." *American Quarterly,* vol. 69, no. 4, 2017, pp. 783–90, https://doi.org/10.1353/aq.2017.0064.

TallBear, Kim. "Beyond the Life/Not-Life Binary: A Feminist-Indigenous Reading of Cryopreservation, Interspecies Thinking, and the New Materialisms." *Cryopolitics: Frozen Life in a Melting World,* edited by Joanna Radin and Emma Kowal, MIT P, 2017, pp. 179–202.

Teperman, Johnny. "Día Mundial de la Poesía se celebrará con poesía chilena y mapuche en Providencia." *BiobioChile.cl,* 18 Mar. 2013, https://www.biobiochile.cl/noticias/2013/03/18/dia-mundial-de-la-poesia-se-celebrara-con-poesia-chilena-y-mapuche-en-providencia.shtml.

"Trutrukero." YouTube, uploaded by Emilio Uribe, 21 May 2009, https://www.youtube.com/watch?v=MzN6J_kGLmo.

Vásquez, Ana. "Expedientes del dolor: mujeres Mapuche en la frontera de la violencia (1900–1950)." *Awükan ka kuxankan zugu wajmapu mew. Violencias coloniales en Wajmapu,* edited by Enrique Antileo Baeza et al., Ediciones Comunidad de Historia Mapuche, 2015, pp. 141–58.

Wetruwe Mapuche. "Quiénes somos." Wetruwe Mapuche, https://wetruwemapuche.wordpress.com/about/. Accessed 15 Dec. 2022.

Zapata, Claudia. *Crisis del multiculturalismo en América Latina. Conflictividad social y respuestas críticas desde el pensamiento político indígena.* Calas and Bielefeld U P, 2019.

2

The Art of an Ecological Constitution

Beyond Derangement in Chile

PAUL R. MERCHANT

On October 14, 2020, a group of women in black mourning clothes processed to the waterfront in the Chilean port of Valparaíso, where they held a symbolic funeral for the Chilean constitution, copies of which were thrown into a boat alongside fragments of text indicating ideas and social norms that the group wished to consign to the past (among them *patriarcado* and *machismo*).[1] The act was accompanied by chants such as "Sin libertad, sin igualdad, no hay derechos ni dignidad. Regresa por donde viniste. Hoy, hundimos el miedo" (Without freedom, without equality, there are no rights or dignity. Go back to where you came from. Today, we drown fear). The boat then left the pier and made its way out to sea, where, implicitly if perhaps not physically, the offending texts were consigned to the deep.

This performance was the work of LasTesis, a Valparaíso-based feminist group which shot to international prominence in November 2019 with *Un violador en tu camino*, a street performance addressing the complicity of the state and society in perpetuating rape culture (for an analysis of the transnational impact of this performance, see Martin and Shaw). This action was just one of many artistic interventions in Chilean public space since the *estallido social*, or social uprising, of October–November 2019. The immediate causes of this extraordinary social movement, which led to a Constitutional Convention and a proposed replacement for the current constitution (which dates from the dictatorship of Augusto Pinochet), might not appear at first glance to be related to the climate crisis or other ecological issues. The rejection of the proposed new constitution in September 2022, moreover, leaves the path forward unclear.[2] This chapter argues, nonetheless, that the ecological focus of many artistic initiatives that emerged in the run-up to or as part of the social movement toward a new constitution merits close attention, not least because these initiatives challenge the peripheral status accorded to ecological concerns in the current constitutional framework, and suggest a more holistic view of human entanglement with the nonhuman world. In interventions by the artistic collective Delight Lab, and in works such as the documentary *Alas de mar* (Mülchi) and *Archipiélago* (Godoy), a questioning of hegemonic modes of aesthetic representation in the context of ecological crisis can be viewed as a call for a new, ecological constitutional settlement that may yet be answered.

Constitutional Derangement

A focus on artworks and artistic interventions that call into question the possibility of representation might seem to sit awkwardly with the contention that the same works point their viewers toward the possibility of a new, ecological constitutional framework. After all, do not all (democratic) constitutional processes and structures entail a degree of representation? Bruno Latour engages with questions of representation in his influential positing of a new Constitution, to replace the modernist Constitution that "saw debates over ecology merely as a mixture to be purified, a mixture combining rationality and irrationality, nature and artifice, objectivity and subjectivity" (129). Latour's definition of representation is, however, characteristically idiosyncratic: in the glossary provided for *Politics of Nature*, two senses of the word are provided. The first is "one of the two powers of (political) epistemology which forbids all public life, since subjects or cultures have access only to secondary qualities and never to essences." The second is "the dynamics of the collective which is re-presenting, that is, presenting again, the questions of the common world, and is constantly testing the faithfulness of the reconsideration" (248).

Latour's constitutional proposal is, as the capitalized orthography of his Constitution suggests, more concerned with the high-level, metaphysical organization of knowledge and life than with the possibility of specific, located constitutional reforms. The two "houses" described in his proposed constitution are given the daunting task of composing a common world, and Latour makes clear that he regards the terminology of parliamentary democracy as "outdated" and simply playing the role of "a white flag waved in the wind so that we can finally negotiate" (165). The simultaneous specificity and abstraction of Latour's proposal, then, make it of limited use for an analysis of a specific case of potential constitutional change in Chile. It is striking, moreover, that among the human actors that Latour identifies as crucial for the construction of a new constitution, there is no mention of artists or creative practitioners, though the constitution itself is described as a work of art. This task is left to scientists, politicians, economists, and moralists (161–63).

This chapter argues, conversely, that creative work, especially when it tests conventional categories of aesthetic representation, is precisely the kind of activity that can afford the formation of better, more inclusive collectives of humans and nonhumans. It argues, moreover, that such work can be seen as a civic endeavor, consonant with the constituent process currently underway in Chile. In that sense, I am advocating, in a specifically ecological arena, the notion of art as a civic agent put forward by Doris Sommer. For Sommer, works of art "on grand and small scales" can "morph into institutional innovation" (3). To illustrate this notion, Sommer gives the now famous examples of Antanas Mockus's use of pantomime artists to regulate traffic in 1990s Bogotá, and Augusto Boal's development of "legislative theater" to improve inclusion and conflict resolution in the city government of Rio de Janeiro (2). How, though, to assess the relation between cultural agency in a broad public sphere and the mechanics of the creation of a new constitution?

This is where an at least partly Latourian approach can be productive. Kerry Whiteside assesses the ability of Latour's proposals to inform "change in the actual constitutional structure of representative democracies," arguing that Latour's proposal that "*every* reflection on governance must take 'nature' into account" can act as a powerful challenge to current democratic practices. Whiteside notes, however, that any actual processes of political-ecological change in the constitutional arena is likely to rely on concepts of agency, deliberation, and reasoning that would qualify for the pejorative adjective "Modernist" in Latour's thought (Whiteside 200, 202). For Whiteside, Latour is mistaken to view the environmental crisis as a failure of the scope of representation—the problem is not that nonhumans have no effective representatives, and the solution is not simply to expand representation. The problem is, rather, the current representational paradigm itself, from which, in this account, Latour never quite detaches his proposals. What is needed, Whiteside argues, is precisely a *constitutional* process, one that determines new paradigms for what counts as representation, because "republics as we know them have stabilized modes of representation that consistently fail to take 'Nature' into account to a degree commensurate with the gravity of the unprecedented, world-altering phenomena that we have collectively unleashed" (203).

At this point, Whiteside's argument comes rather close to that advanced by Amitav Ghosh in *The Great Derangement,* his analysis of the modern novel's inability fully to deal with the scale and nature of the effects of climate change. This chapter's subtitle is "beyond derangement" because the artistic interventions that I'm going to discuss offer some hope that Ghosh's diagnosis of a "great derangement" in the modern novel's attitude to the climate crisis need not extend to other forms of aesthetic action. Whereas the realist literary novel, for Ghosh, traffics in probabilities that obscure the (improbable) real, and in "settings" that are disconnected from global networks (59), the artworks to be discussed here ask their spectators to look again at their apparently banal everyday environments and to see, hear, and feel them anew. These are works that often resist containment within one medium, and that engage their spectators on multiple fronts—they are multisensory experiences. And they are precisely concerned with the ways in which experience exceeds conventional structures of *representation.* We might think of them, then, as an aesthetic response to Whiteside's call to "make good the defective ability of existing representative institutions to take up concern for 'nature'" (200).

A focus on Chile may nonetheless seem a stubbornly national framework for thinking about global ecological concerns, but I think it is possible to argue, pace Ghosh's skepticism on the utility and fairness of the nation-state, which he views as a correlate of the novel (59), that the nation can be a framework for an honest and thorough engagement with the realities of the climate crisis. In the artworks discussed below, the nation is at least implicitly posed as a framework for understanding the problem, and for imagining potential solutions. The fact that the management and conservation of bodies of water, from rivers and glaciers to the Pacific Ocean, emerge as key issues of concern means that the nation cannot be the ultimate horizon of this discussion, but it is perhaps a starting point. In the action by LasTesis with which this chapter began, for example, the ocean is figured as a space for renewing the constitutional framework of the nation.

Before moving to consider the artworks in detail, though, it is important to have a clear sense of the existing constitutional architecture and the ways in which possible changes to it have been sketched out. The current Chilean constitution contains only one substantive reference to environmental matters: Article 19 (Section 8) guarantees "el derecho a vivir en un medio ambiente libre de contaminación" [the right to live in an environment free from pollution] for all and goes on to state that "es deber del Estado velar para que este derecho no sea afectado y tutelar la preservación de la naturaleza [it is the duty of the State to ensure that this right is not affected, and to safeguard the preservation of nature]" (Constitución Política de la República de Chile).[3] This promised right has not, however, proved enforceable, as the proliferation of highly polluted "sacrifice zones" demonstrates (Ramírez Nova). By contrast, the 2022 constitution proposal included an entire chapter dedicated to nature and the environment, and the first article in that chapter (no. 127) states that "la naturaleza tiene derechos [nature has rights]" (*Propuesta de Constitución Política*, 45). The chapter goes on to set out a duty for the state to protect biodiversity, establishes a category of "bienes comunes naturales" (natural common goods) with special protections for water, and creates a body charged with protecting the rights of nature—the Defensoría de la Naturaleza.

The proposed constitution thus went a long way toward implementing ideas advocated by many activists and scholars of political ecology. One prominent member of the latter group, Eduardo Gudynas, had suggested to the Constitutional Convention's commission on the environment that a new constitution, rather than determining an economic model for the nation, should consider what kinds of value nature possesses, beyond the economic (as realized in Chile's current extractivist economic model). These might include, Gudynas argued, aesthetic and historical value, as well as the value ascribed to nature by Indigenous peoples, and a concept of "simples valores ecológicos" (simple ecological values). A recognition of these kinds of value at a constitutional level would then, Gudynas suggested, necessitate a constitutional guarantee of the rights of nature (Comisión sobre Medio Ambiente, Derechos de la Naturaleza, Bienes Naturales Comunes y Modelo Económico). In what follows, I argue that the challenges to representation, or expanded forms of representation, of the ecological artwork emerging alongside and from within Chile's recent social movements encourages a reframing (whether legal or otherwise) of human relations with "nature," in a manner consonant with Gudynas's submission to the Convention's environmental commission.

Delight Lab: Projecting Alternative Futures

The Chilean *estallido social* provoked an explosion of politically oriented creativity, or "artivism," to use a term with Latino/a and Chicano/a roots. Yet, with the exception of LasTesis, perhaps no group has become more publicly associated with the *estallido* than Delight Lab, an experimental audiovisual design studio whose projections on the Torre Telefónica in Santiago de Chile's Plaza de la Dignidad (known before the estallido as Plaza Italia) have become emblematic instances of contemporary protest aesthetics. The projections have, moreover, generated significant repercussions within existing institutional structures. To give just one example: on 19 May 2020, while Delight Lab were projecting the word "HUMANIDAD" (HUMANITY) on the Torre Telefónica, an unmarked truck protected by members of the Carabineros (Chile's principal national police force) used its headlights to block out the projection. After a legal action in the Chilean courts was unsuccessful, Delight Lab appealed to the Inter-American Commission on Human Rights, alleging censorship of their work ("Delight Lab recurrirá ante la CIDH tras rechazo de la Corte Suprema a reconocer censura en su contra"). The collective has also taken an explicit interest in the new constitutional process, projecting the word "RENACE" ["IS REBORN"] when there was a decisive national vote for a new constitution in later October 2020.

Beyond the immediate context of the estallido and its aftermath, much of Delight Lab's work has linked a revaluation of Indigenous knowledges and peoples to an enhanced ecological sensibility. María José Barros has documented interventions carried out in collaboration with Corporación Traitraico in the Chilean part of Wallmapu, the ancestral territory of the Mapuche people. Barros argues that the projection of images on the banks of the Rawe and Pilmaiken rivers, protesting the planned construction of hydroelectric dams by a Norwegian corporation, contrasts a Mapuche spiritual understanding of rivers as the means by which the souls of the dead return to the sea with the Chilean state's willingness for water rights to be bought and sold like any other commodity (Barros). In the most recent edition of the Santiago a Mil theater festival, the collective presented *Espíritu del agua*, in January 2021, a series of animations projected on water towers across Santiago and in Concepción, a city in the south of Chile. *Espíritu del agua* begins and ends with the following incantation in Mapudungun, performed by the Mapuche Werken (cultural elder) Joel Maripil, which puts forward a vision of human existence as inherently fluid: "Del agua salimos y al agua volveremos. Hemos sido río, vapor y hielo. Hemos andado en corrientes que suben y bajan, que se hunden en la tierra o se disuelven en el mar. El océano es el gran espíritu del que todo emerge y al que todo regresa; somos fragmentos de él, lágrimas de gozo o tristeza que salen en cuerpos y regresan en espíritu [From water we came and to water we will return. We have been river, vapor, and ice. We have moved in currents that rise and fall, which sink into the ground or dissolve in the sea. The ocean is the great spirit from which all emerges and to which all returns; we are fragments of it, tears of joy or sadness that go out in bodies and return as spirit]" (Delight Lab).

Here, we find an identification of human embodiment with water that recalls Astrida Neimanis's conception of humans' watery embodiment and Stacy Alaimo's notion of "trans-corporeality at sea." This notion of fluid and mutable corporeality allows a call for the conservation of bodies of water to be rendered as a call for the preservation of human souls: "Las almas regresan al mar a través de los ríos. Pero así como los selk'nam, los ríos viven amenazados por la codicia de las personas desconectadas de la tierra [Souls return to the sea through rivers. But like the Selk'nam, rivers are threatened by the avarice of people who are disconnected from the land]" (Delight Lab). Moreover, the projection of crashing ocean waves on a concrete water tower, and the prevalence of fade cuts and close-ups of flowing water in the audiovisual montage, ironically relegate the infrastructure of modern water management, with its insistence on containment and separation, to the status of mere background for the envisioning of a different, more fluid relation between humans and their environment. The projections also reclaim the infrastructure of a privatized water management system as a resolutely public forum for experimental forms of aesthetic representation.

It is possible to imagine a critique of Delight Lab's interventions along the lines of that advanced by David Chandler and Julian Reid in their excoriating analysis of the "speculative turn" in anthropology: for Chandler and Reid, the work of Eduardo Viveiros de Castro, Marisol de la Cadena, and others seeks to "distil Indigenous knowledge as method or analytics that allows the development of a new branch of speculative philosophy" (498). It should be acknowledged, of course, that Andrea and Germán Gana Muñoz, the siblings who make up Delight Lab, are not themselves of Indigenous descent, and so it might be said that their work makes use of Indigenous concepts in the service of their own creative ends. Yet the fact that many of their interventions have taken place in Wallmapu, in collaboration with Mapuche actors (such as Corporación Traitraico and Joel Maripil) and were designed to raise awareness of political and ecological realities affecting the lives of Mapuche communities, suggests that that critique would not be entirely fair in this case. The importance of political context to the works produced by Delight Lab can be seen, for instance, in the timing of *Espíritu del agua,* which was presented in the run-up to elections for Chile's Constitutional Convention, held in May 2021. It has therefore been seen as a call for the new constitution to abandon the privatized water rights of the 1981 Water Code, an offshoot of the 1980 constitution that is still in force today (on the uniquely privatized nature of post-1980 water management in Chile, see Budds).

It is far from the only such call. To give just one further example: in September 2020, the ecological magazine *Revista Endémico* issued a call for "posters for an ecological constitution"—posters designed to encourage the Chilean public to vote for candidates for the Constitutional Convention who espoused ecologically conscious positions. Launching the call, the director of *Endémico*, Nicole Ellena, noted that there was a long tradition of graphical activism in Chile, and offered a view of art and design as "herramientas de cambio, [cuyo] rol es clave para darle una voz a nuestro planeta y sus habitantes no humanos [tools of change, [whose] role is key to giving a voice to our planet and its nonhuman inhabitants]" ("Convocatoria abierta para campaña 'Carteles para una constitución ecológica'"). Ellena's phrasing here clearly allocates the (envisaged) posters a representative role, not only in aesthetic terms but also politically, and thus recalls the debate around Latour's propositions discussed above. The posters that were eventually selected by *Endémico*'s editorial team, meanwhile, are heterogeneous and impossible to fully categorize as a group, though there are some common features across the individual works. These include the use of bright colors, clear lines, and bold block text (all of which indicate, of course, the close relation of these works to posters and placards carried in the protests that took place between October 2019 and early 2020). Several of the selected posters include no text beyond the phrase *constitución ecológica* (ecological constitution) itself, while others opt for short slogans, such as "De la tierra venimos, a sus entrañas volveremos" (We come from the earth, and we will return to its entrails) and "La ecología es más importante que cualquier ideología" [Ecology is more important than any ideology]. One poster reworks Chile's national motto, "Por la razón o la fuerza" (By reason or strength), into "Ni por la razón ni por la fuerza / por la fuerza de la naturaleza" (Neither by reason nor by strength / by the strength of nature). What these three examples have in common is a ceding of protagonism to a nonhuman actor, variously named as the earth, ecology, or nature.

Here, again, we witness a point of contact with the 2022 constitutional proposal's establishment of "nature" as a subject of rights. There is a connection with Delight Lab's work, too, in that *Endémico*'s initiative demonstrates a commitment to art's place in the public sphere. It is perhaps misleading to refer to a singular "public sphere," however, as the posters are available in an open-access virtual exhibition on the *Endémico* website, as well as on

Endémico's social media channels, but have also been exhibited in physical spaces, such as in the Parque Cultural in Valparaíso in June 2021.

The form of the interventions described above, which combine written or spoken language with audiovisual media, and moreover use language in a way that exceeds simple indication of a signified object, mean that they can be viewed as examples of an "ecological avant-garde" in Chile.[4] Where Ghosh argues that the novel "always align[s] itself with the avant-garde as it hurtles forward in its impatience to erase every archaic reminder of Man's kinship with the nonhuman" (70), a perspective from Chile offers a more complex conception of avant-garde practice. Somewhat paradoxically, a work like *Espíritu del agua* stages a rupture in our typical modes of relating to our surroundings, transforming a water tower into a projection screen, but orients this rupture in experience precisely toward a reminder of "kinship," in Ghosh's terms, between humans and their nonhuman environs. I have already briefly suggested that it is no accident that this project makes use of watery environments in suggesting this connection. In what follows, I expand on this observation to suggest that an "archipelagic" reading of cultural production might offer a way of imagining an inclusive ecological constitution.

An Archipelagic Constitution?

The notion of "archipelagic thought" is rather more familiar in Caribbean and Pacific Island contexts than in studies of the cultures of the Southern Cone: through the work of Édouard Glissant, for instance, or in Epeli Hau'ofa's now-canonical notion of the "sea of islands," which values islander knowledge and relations with marine and island environments above the colonizer's view of the ocean as the empty space in between: "There is a world of difference between viewing the Pacific as 'islands in a far sea' and as 'a sea of islands.' The first emphasizes dry surfaces in a vast ocean far from the centers of power. Focusing in this way stresses the smallness and remoteness of the islands. The second is a more holistic perspective in which things are seen in the totality of their relationships" (152–53).

Craig Santos Perez has recently drawn on Hau'ofa's groundbreaking work in his analysis of the diasporic cultural identity of Chamoru islanders. And, beyond Oceania, Brian Russell Roberts and Michelle Ann Stephens have suggested that an "archipelagic American studies" can offer a way of "de-continentalizing" our understandings of space and identity. A way, in other words, of recognizing the cultural and political value of apparently marginal or in-between spaces like islands, seas, beaches, and inlets, and the people who live in them. Jonathan Pugh and David Chandler, meanwhile, argue in *Anthropocene Islands* that "work with islands has become productive in the development of many of the core conceptual frameworks of Anthropocene thinking" (2). This is in part because of their frequent position on the front lines of environmental change (rising sea levels), but also because of how islands often come to constitute distinct but related ontological units, for their inhabitants and also for those who visit and study them. In that respect, they are ideal places for thinking through the ever more complex relational entanglements between humans and nonhumans, and indeed between humans and other humans, that characterize our contemporary world.

Chile is, in geographic terms, unquestionably an archipelagic nation: one need only look at a map to establish that. South of Puerto Montt, the land fragments into hundreds of islands.[5] Through an analysis of some recent documentary films and a multimedia sound art/installation project, I will argue here that a decontinental or archipelagic understanding of the Chilean nation, which foregrounds shared exposure to and entanglement with the world, might provide a productive basis for the elaboration of a truly inclusive ecological constitution.

In Patricio Guzmán's documentary *El botón de nácar*, we meet Martín González Calderón, a Yaghan man from Tierra del Fuego who explains how the Chilean Navy's strict control over maritime space has made it almost impossible for him and his family to travel by boat using the skills and techniques passed down over generations. Guzmán also speaks to Gabriela Paterito, a Kawésqar woman who recounts a long journey by canoe that she made when she was a girl, and the director prompts her to state that she does not feel Chilean at all. In Guzmán's film, Indigenous mobility by water in the Patagonian archipelago is presented as lost to the past, and impossible in the present.[6] Other filmmakers have taken a different approach to these issues, however. In 2016's *Tánana, estar listo para zarpar*, for instance, we meet Martín González Calderón again, but this time at much greater length. The documentary's directors Alberto Serrano Fillol and Cristóbal Azócar do not provide an explanatory voiceover. Instead, the camera follows González Calderón as he goes about his daily life, and then seeks to build a boat in which he can re-create a childhood trip around the False Cape Horn, near the southern tip of the continent, that he undertook with his father. González Calderón explains the intertwining of his memories and the landscapes and seascapes of the archipelago in fragments, and on his own terms: local ecological knowledge is not offered up to the viewer for easy consumption in this case.

Another documentary from 2016, *Alas de mar,* exhibits some similar characteristics. Here, the director Hans Mülchi does provide a voiceover, but it is intermittent and reflective. The film follows the journey by boat of two Kawésqar women, Rosa and Celina, back to the region where they grew up. The voices of Rosa and Celina are much more prominent than that of Mülchi, or indeed that of the European anthropologist who is traveling with them. It is not only the human voice that counts, though. Both *Tánana* and *Alas de mar* contain long sequences in which the only sounds audible are the sounds of travel by sea: the flapping of a sail, the rush of the wind, the crash of waves against the hull, or the roar of a motor. This openness to the sounds of the marine environment allows the spectator to share in the embodied experience of the protagonists in a way that escapes any definitions that might be imposed by spoken or written language. It is precisely because *Alas de mar* and *Tánana* do not offer definitive answers to the question of the relation between Indigenous identity and Chilean identity that I find them valuable to think with. The people whose stories are told in these films have been displaced from their childhood homes (as is the case for Rosa and Celina) or are held in place by the state's unwillingness to allow maritime travel outside of specific, limited purposes (in the case of Martín). And yet we see them strive to retrace past journeys and reclaim certain modes of mobility as an essential part of their heritage. In fact, Indigenous identity itself appears as fluid and mobile in these films. Martín notes that while he understands much of the Yaghan language, he cannot speak it well himself, and in *Tánana* we see him teaching boatbuilding techniques to family members who are clearly of mixed heritage. In *Alas de mar,* Rosa and Celina share weaving and construction techniques with their fellow travelers.

These films' acts of representation, their visions of mobile and changing identities, present a source of inspiration for a plurivocal or even plurinational political order, of the kind that was represented by the formalized participation of many Indigenous groups, including the Kawésqar and the Yaghan, in the Constitutional Convention, and eventually codified in the 2022 proposal, which describes Chile as "plurinational, intercultural, regional and ecological" (*Propuesta,* 5).[7] The fact that 61.89 percent of voters rejected this vision of the country in September 2022 makes it evident that there is still much work to be done in communicating the value of such visions. This much was becoming clear even before the 2022 referendum: the house of the Kawésqar representative at the Constitutional Convention, Margarita

Vargas, was burned to the ground in October 2021, though no one was harmed (Hermosilla).

Might there still be room in an eventual new constitution for the varied "ecological epistemes" on display in these films (Escobar 62, quoting Leff, La apuesta por la vida)? Paradoxically, perhaps, the failure of the 2022 constitutional proposal makes the persistence of cultural objects that articulate such ideas even more important, even when the very limited distribution circuits for documentary film in Chile restrict the extent to which the works discussed above can be seen to act in the modes that Sommer envisions in her articulation of cultural agency.

It is for this reason that I turn, as a form of provisional conclusion, to a multimedia project titled *Archipiélago*, which began in 2018 and has evolved across several platforms and formats. In 2018, the artists Fernando Godoy, Esteban Agosín, and Carlos Lértora hired a boat and undertook three journeys of four to ten days each around the archipelago of Chiloé. During these journeys, the artists lived on the boat, which they thought of as a kind of laboratory, and used hydrophones and other recording equipment to capture the sounds of the underwater seascape around the islands, paying particular attention to how the sounds of human activity (e.g., through salmon farming) were interfering with sounds produced by other actors within the ecosystem. Godoy has stated in an interview that the group set out to understand the impact of "extractivism in the sea," in the form of the extensive salmon farming that takes place in the south of Chile, by exploring its sonic effects (*Panel 1*). In addition to recording the sounds, Godoy and his collaborators also turned the boat into a broadcaster, sharing the underwater seascapes online via radiotsonami.org.

However, in Godoy's account of the project, this activity soon proved unsatisfactory, and the group of artists realized that in order to make sense of the sounds they were recording, they needed to turn to local residents and record the stories of those who had lived on and interacted with the ocean, whose way of life was on the brink of disappearing. This content, alongside the underwater sounds, was then also transmitted via FM radio, so that it could reach local communities (*Panel 1*). The documentation of the project includes a documentary directed by Lértora (2019), an installation at the Museo de Arte Moderno Chiloé in 2019, and a book (Godoy). Thus *Archipiélago*, which started in response to a particular ecological disaster—the *marea roja*, or toxic algal bloom, of 2016, which may have been caused by the dumping of dead salmon (Armijo et al.)—has evolved in relation to the people and environments with which it has interacted. The project is both singular and multiple and seeks a variety of forms of representation. These two facts alone render *Archipiélago* a helpful model for considering the possible forms of a new ecological constitution. Godoy, Lértora, and Agosín's project illustrates why the archipelago as lived and imagined space is a useful conceptual tool for articulating the complexity of human-nonhuman (and human-human) relations in contemporary Chile. As Yolanda Martínez-San Miguel and Michelle Stephens put it: "The archipelago calls for a meaning-making and rearticulation that responds to human experiences traversing space and time. Archipelagoes happen, congeal, take place. They are not immanent or natural categories existing independently of interpretation. Yet they can also become an episteme, an imaginary, a way of thinking, a poetic, a hermeneutic, a method of inquiry, a system of relations" (3).

This conception of the archipelago as a plurivalent episteme is consistent with Gudynas's call for a new Chilean constitution to embrace, or at least engage with, multiple values for nature. Writing elsewhere, Gudynas has argued that "the recognition of plural valuations, including intrinsic values in non-humans, is an openness to other sensitivities and practices that generate different moral mandates, public policies, understandings of justice" (241). The aftermath of 2019's estallido social has to date seen two failed attempts to generate consensus around a new set of constitutional principles, in 2022 and 2023. A question thus remains: How can artistic interventions of the kind examined here generate such "openness to other sensitivities"? Might we imagine a Chilean variant of Boal's "legislative theater" (Sommer 50–60), a kind of "constitutional theater" embracing an archipelagic vision of the nation, as a mechanism for achieving a still-elusive consensus on new shared values within a constituent process? There is no guarantee of success, but where constituent assemblies, conventions, and councils have failed, perhaps creative methods should be given a chance to take on Latour's challenge of composing a common world.

Notes

1 An earlier version of this chapter was published as Merchant, Paul. "Constitutional and Ecological Entanglements in Contemporary Chile." *Stories Come to Matter: Water, Food, and Other Entanglements,* edited by Santiago Alarcón-Tobón and Enric Bou, Edizioni Ca' Foscari, 2025, pp. 27–42.
2 At the time of writing (May 2025), the July 2022 constitutional draft had been rejected in a plebiscite, as had a second, much more conservative draft (in a referendum in December 2023). The government of President Gabriel Boric has not since proposed any projects of large-scale constitutional reform.

3   Unless otherwise indicated, all translations are my own.
4   For an analysis of the history and present of ecological avant-garde practices in Chile, see Merchant ("Cecilia Vicuña's Liquid Indigeneity").
5   This is without even considering Rapa Nui (Easter Island), the island some four thousand kilometers from the South American continent that is administered by Chile.
6   For a more detailed discussion of how Guzmán consistently relegates Indigenous experience to a separate time frame, or even a separate world, see Merchant ("'Collecting What the Sea Gives Back'").
7   "Plurinacional, intercultural, regional y ecológico."

Works Cited

Alaimo, Stacy. "States of Suspension: Trans-Corporeality at Sea." *ISLE: Interdisciplinary Studies in Literature and Environment*, vol. 19, no. 3, Dec. 2012, pp. 476–93. Silverchair, https://doi.org/10.1093/isle/iss068.

Alas de mar. Directed by Hans Mülchi, Blume Producciones, 2016.

Armijo, Julien, et al. "The 2016 Red Tide Crisis in Southern Chile: Possible Influence of the Mass Oceanic Dumping of Dead Salmons." *Marine Pollution Bulletin*, vol. 150, Jan. 2020, p. 110603. ScienceDirect, https://doi.org/10.1016/j.marpolbul.2019.110603.

Barros, María José. "Delight Lab en Wallmapu: Descolonizando la mirada y las aguas." Revista *Endémico*, 15 July 2020, https://www.endemico.org/delight-lab-wallmapu-descolonizar-la-mirada-las-aguas/.

Budds, Jessica. "Power, Nature, and Neoliberalism: The Political Ecology of Water in Chile." *Singapore Journal of Tropical Geography*, vol. 25, no. 3, Nov. 2004, pp. 322–42. Wiley Online Library, https://doi.org/10.1111/j.0129-7619.2004.00189.x.

Chandler, David, and Julian Reid. "Becoming Indigenous: The 'Speculative Turn' in Anthropology and the (Re)Colonisation of Indigeneity." *Postcolonial Studies*, vol. 23, no. 4, pp. 485–504. Zotero, https://doi.org/10.1080/13688790.2020.1745993.

Chile. Constitución Política de la República de Chile. Gobierno de Chile, 2024. https://www.camara.cl/camara/doc/leyes_normas/constitucion.pdf. Accessed 20 November 2025.

Comisión Sobre Medio Ambiente, Derechos De La Naturaleza, Bienes Naturales Comunes Y Modelo Económico. Sesión N° 8 del jueves 4 de noviembre de 2021, https://www.cconstituyente.cl/comisiones/verDoc.aspx?prmID=714&prmTipo=DOCUMENTO_COMISION.

"Convocatoria abierta para campaña 'Carteles para una constitución ecológica.'" *Revista Endémico*, 28 Sept. 2020, https://www.endemico.org/carteles-para-una-constitucion-ecologica/.

Delight Lab. "*Espíritu del agua*." Santiago a Mil 2021, https://www.santiagoamil.cl/obras-2021/espiritu-del-agua/. Accessed 2 Mar 2021.

"Delight Lab recurrirá ante la CIDH tras rechazo de la Corte Suprema a reconocer censura en su contra." *El Mostrador,* 29 April 2021, https://www.elmostrador.cl/cultura/2021/04/29/delight-lab-recurrira-ante-la-cidh-tras-rechazo-de-la-corte-suprema-a-reconocer-censura-en-su-contra/.

El botón de nácar. Directed by Patricio Guzmán, Atacama Productions, Valdivia Film, Mediapro, 2015.

Escobar, Arturo. *Pluriversal Politics: The Real and the Possible.* Duke UP, 2020.

Ghosh, Amitav. *The Great Derangement: Climate Change and the Unthinkable.* U of Chicago P, 2017.

Godoy, Fernando. *Archipiélago.* Tsonami Ediciones, 2023.

Gudynas, Eduardo. "Value, Growth, Development: South American Lessons for a New Ecopolitics." *Capitalism Nature Socialism,* vol. 30, no. 2, Apr. 2019, pp. 234–43. Taylor and Francis+NEJM, https://doi.org/10.1080/10455752.2017.1372502.

Hau'ofa, Epeli. "Our Sea of Islands." *Contemporary Pacific,* vol. 6, no. 1, 1994, pp. 148–61.

Hermosilla, Ignacio. "Incendio destruye vivienda de convencional Kawésqar en Magallanes." BioBioChile, 25 Oct 2021, https://www.biobiochile.cl/noticias/nacional/region-de-magallanes/2021/10/25/margarita-virginia-vargas-convencional-pide-que-se-investigue-incendio.shtml.

Latour, Bruno. *Politics of Nature: How to Bring the Sciences into Democracy.* Translated by Catherine Porter, Harvard UP, 2004.

Martin, Deborah, and Deborah Shaw. "Chilean and Transnational Performances of Disobedience: LasTesis and the Phenomenon of 'Un Violador En Tu Camino.'" *Bulletin of Latin American Research.* Wiley Online Library, https://doi.org/10.1111/blar.13215. Accessed 25 Feb. 2021.

Martínez-San Miguel, Yolanda, and Michelle Stephens. "'Isolated Above, but Connected Below': Toward New, Global, Archipelagic Linkages." *Contemporary Archipelagic Thinking: Toward New Comparative Methodologies and Disciplinary Formations,* edited by Michelle Stephens and Yolanda Martínez-San Miguel, Rowman & Littlefield, 2020, pp. 1–44.

Merchant, Paul. "Cecilia Vicuña's Liquid Indigeneity." *Liquid Ecologies in Latin American and Caribbean Art,* edited by Lisa Blackmore and Liliana Gómez, Routledge, 2020, pp. 188–206.

Merchant, Paul. "'Collecting What the Sea Gives Back': Postcolonial Ecologies of the Ocean in Contemporary Chilean Film." *Bulletin of Latin American Research,* vol. 41, no. 2, 2022, pp. 209–26. Wiley Online Library, https://doi.org/10.1111/blar.13231.

Neimanis, Astrida. *Bodies of Water: Posthuman Feminist Phenomenology.* Bloomsbury Academic, 2017.

Panel 1: Arte y ecología en Chile. Directed by Fundación Mar Adentro, 2021. YouTube, https://www.youtube.com/watch?v=YKgwGm4Q-3Q.

Propuesta de Constitución Política de la República de Chile. LOM Ediciones, 2022.

Perez, Craig Santos. "'The Ocean in Us': Navigating the Blue Humanities and Diasporic Chamoru Poetry." *Humanities*, vol. 9, no. 3, 3, Sept. 2020, p. 66. www.mdpi.com, https://doi.org/10.3390/h9030066.

Pugh, Jonathan, and David Chandler. "Anthropocene Islands: Entangled Worlds." U of Westminster P, 2021. www.uwestminsterpress.co.uk, https://doi.org/10.16997/book52.

Ramírez Nova, Matías. "Crisis Socioambiental Y Zonas De Sacrificio: *Propuesta*s Para Una Constitución Ecológica." *Revista Debates Jurídicos y Sociales*, no. 7, 2020, pp. 93–104.

Roberts, Brian Russell, and Michelle Ann Stephens, editors. *Archipelagic American Studies*. Duke UP, 2017. Open WorldCat, https://dx.doi.org/10.1215/9780822373209.

Sommer, Doris. *The Work of Art in the World: Civic Agency and Public Humanities*. Duke UP, 2014.

Tánana, estar listo para zarpar. Directed by Alberto Serrano and Cristóbal Azócar, Tecleo producciones, 2016.

Whiteside, Kerry H. "A Representative Politics of Nature? Bruno Latour on Collectives and Constitutions." *Contemporary Political Theory*, vol. 12, no. 3, Aug. 2013, pp. 185–205. Springer Link, https://doi.org/10.1057/cpt.2012.24.

3

Entangled Ruins

Nature and Postcatastrophic Landscapes in
Hiram Bingham's *Inca Land* and José de la Riva-Agüero's
Paisajes peruanos

ANDRÉS ERNESTO OBANDO

Recent years have seen an extraordinary interest in fictions that describe apocalyptic and postapocalyptic landscapes. Although it is not a new trend, the peculiarity of many of these narratives about the end of human civilization is that they feed on contemporary narratives of environmental catastrophes. Thus, in the face of the imminent possibility of the extinction of a critical number of species and ecosystems that define the Anthropocene era, it is not at all surprising that narrative and visual forms emerge to imagine *that* future.

Among all the visions that have come to be common in books, movies, or even video games, there is one that is particularly striking. In this postapocalyptic scenario, the end of human civilization is represented through overgrown buildings, streets populated with wild animals, cities reduced to ruins in which the debris of the past mixes with an expansive, uncontrollable but, in any case, growing vegetation. Overall, this is intended to show that human civilization has fallen. And yet this is not the most typical way of understanding the zeitgeist of the twenty-first century. As James Painter's study shows, more than 80 percent of climate change stories written between 2010 and 2012 in major newspapers of countries such as Australia, France, India, the United States, or England employed a "disaster frame" where species death, ice melting, and human extinction, among other topics, were the norm (57–78). Despite the prevalence of this "disaster frame" in contemporary journalism, narratives about a flourishing nature reclaiming what was once a thriving city have also been a long-standing part of the many imagined futures for the planet.

It is undeniable that a certain green optimism emanates from the perspective of an intertwining of nature and the built environment, one that has even given rise to diverse approaches such as urban ecology, regenerative architecture, or neoromantic proposals like Jonk's "Natura." Likewise, scholars such as Felix Kirschbacher or Eva Horn have emphasized that images of a revitalized nature, usually represented through lush vegetation, connect with the dreams of many rewilding projects and generally serve to appease anxieties regarding the extinction of life that the Anthropocene raises. As Alan Weisman's visions of a world without us show, in the end, life will manage to continue even if humans are no longer present on the planet.

But, behind this somewhat comforting scenario, there is another layer that deserves attention. This is mainly because the fantasy of a planet Earth dominated by fungi, savage megafauna, or fierce tropical plants is not necessarily a happy dream but rather a posthumanist nightmare. And, as such, it should be connected with some of the specters that have haunted visions of humanity's future for a long time.

One way to begin unpacking the anxieties embedded in these narratives of a resurgent nature is by tracing their connections to some of the enduring themes at the core of Western modernity. And there is probably no better place to start than with the very notion of the ruin. Since the future—at least one of the possible futures available to our species—is imagined as a landscape of overgrown ruins, it might be a fertile line of inquiry to explore how modernity has discursively constructed the concept of ruins as part of its vision of imperial decline and civilizational collapse.

But what quickly becomes evident from this line of thought is that descriptions of nature gradually overtaking abandoned buildings—or even entire cities—have long been among the most persistent and recurrent tropes in representations of the ruins of civilizations. So, perhaps, at least in this case, it makes sense to interrogate the ruins of the past in order to understand the ruins of the future.

In this study, I would like to explore this possibility. Instead of examining the ways in which contemporary narratives portray the ruins of the future, I turn to the past to analyze the role that the modern conceptualizations of nature have played in representing a postcatastrophic scenario such as the landscape of ruins that seems to emerge after the fall of an empire or the decline of a civilization. The occasion for this analysis is a paradigmatic case within postcolonial studies: La Sierra Peruana (Peruvian Highlands), but more specifically two examples of modern narratives that gained momentum in the first half of the twentieth century, and in which both nature and ruins play a central role in characterizing this region: Hiram Bingham's *Inca Land: Explorations in the Highlands of Peru* (1922) and José de la Riva-Agüero's *Paisajes peruanos* (Peruvian landscapes, 1955).[1]

As several scholars have noted, both Hiram Bingham and José de la Riva-Agüero were central figures in the consolidation of the South Peruvian Highlands as a differentiated cultural landscape (Beasley-Murray; Vich). Through their essays, letters, or travelogues, they created a powerful narrative in which the Andean highlands were reimagined as the "Inca land" and, consequently, as a landscape primarily characterized by Inca ruins. However, while studies by Flores Galindo, Castro-Klarén, and López Lenci focus on the postcolonial implications of this identification, there is a notable absence of research addressing the role modern ideas about nature played in crafting these narratives. By considering Hiram Bingham's *Inca Land* and José de la Riva-Agüero's *Paisajes peruanos,* I attempt not only to fill this gap but also to show the relevance that nature has in modern narratives about *La Sierra*. My hope is that, through this analysis, we can understand the conception of nature underlying Western modernity's imagining of a civilization's decline, as well as the deep connections this vision has with some of the catastrophic scenarios envisioned as part of the Anthropocene era.

The first two sections examine the role of the concept of nature within the discursive production of archaeological ruins in the Peruvian Andes. Despite the extraordinary variety of examples included in *Inca Land* and *Paisajes peruanos,* I pay special attention to Bingham's version of his first encounter with the ruins of Machu Picchu, and Riva-Agüero's description of the ruins of Vilcashuamán. Although John Beasley-Murray's article "Vilcashuamán: Stories in Ruins" will be an inevitable point of reference in my analysis, here I take a different direction by focusing specifically on the relationship between ruins and nature—mostly vegetation—and not on the posthegemonic dimension of Vilcashuamán's ruins.

The third section explores a broader panorama: the landscape of *La Sierra*. In brief, I claim that to reinforce the idea of *La Sierra* as a postcatastrophic scenario made out of ruins, these writers resorted to detailed descriptions of Andean nature that emphasized its expansive, chaotic, and untamed character. While I acknowledge that these types of narratives contribute to establishing a historical discontinuity between the past and present of the Peruvian Andes, I also argue that, from a more general perspective, they are an illustration of how modernity externalizes its antagonisms (e.g., nature or ruined landscapes) by relocating them at the peripheries. Building on Gaia Giuliani's distinction between "places *for* disaster" and "places *of* disaster" (157), in the final section, I highlight the connections between this sort of externalization of modernity's threats and the dystopian images of a world reduced to overgrown ruins.

Two considerations before beginning. In the first place, the choice of a setting such as *La Sierra* is not accidental. As will become clear later, one of the conclusions that can be drawn from my analysis is that postcatastrophic images of a world in ruins covered by nature replicate, on a planetary scale, some of the elements with which modernity represented ruinous landscapes in its peripheries. This fact, I believe, could be taken as an argument in favor of the planetarization of some of the fears that modernity tried to displace to its margins.

However, I am well aware that this may not be the most orthodox approach to understanding the challenges posed by the Anthropocene. After all, what can these texts—texts that belong to an era when the global ecological disaster was not even considered a possibility—tell us about our future or even our present? Yet perhaps that is precisely the point. In their Introduction to *The Arts of Living in a Damaged Planet,* Anna Tsing et al. write that "the winds of the Anthropocene carry ghosts—the vestiges and signs of past ways of life still charged in the present" (1). And what is more important: that we need to return to these ghosts, to these multiple pasts, to see the present more clearly.

Of course, hauntology is also relevant in this scenario. Building on Jacques Derrida's foundational ideas, Mark Fisher states that "the future is always experienced as a haunting" (16), but also that hauntology is "the way in which the past has a way of using us to repeat itself" (19). This essay gathers these and other insights to appreciate how the ghost of colonialism still haunts our visions of the future. The relevance of the Sierra case, and consequently the work of Bingham and Riva-Agüero, stems precisely from this fact. Through their works, it becomes possible to glimpse how certain ways of understanding nature, framed within colonial relations of dominance, continue to influence some of the futures we imagine and, by extension, our present.

The second comment is a matter of terminology. Throughout the text, I assume the terms postapocalyptic and postcatastrophic to be equivalent. The reason for this is that, as Teresa Heffernan points out, since the nineteenth century, the meaning of terms such as "apocalypse" or "apocalyptic" has begun to shift away from their original religious sense (linked to revelation or emergence of a new world) to a secularized form that relates them to catastrophe or disaster (7). Connected to this shift, James Berger has noted another important transition from apocalyptic narratives (linked to an irremediable end) to postapocalyptic or postcatastrophic narratives—that is, narratives about the remains and ruins that persist after the end, after the "unspeakable trauma" has taken place (19). This chapter departs from these insights and develops them by analyzing a context such as *La Sierra,* where what remained after the catastrophe of colonialism were ruins and nature.

Ruins and "the Embrace of Nature"

In his article "Ruins and the Embrace of Nature," John Dwyer attributes the emotional impact that ruins generate to "the Ozymandias effect"—that is, "the realization that the mightiest works of humans are transient, and may not, in the long run, resist the forces of nature" (10). Dwyer's remarks, however, are mainly aimed at the Global North, specifically at British and Roman ruins. According to Dwyer, in those contexts, the presence of plants, moss, or fungi intensifies the Ozymandias effect, turning more vivid and ineluctable the transitoriness of the human works and contributing to the fetishization or aestheticization of the ruin. However, something slightly different happens in postcolonial contexts. In these scenarios, the embrace of nature—as Dwyer poetically describes the wearing effect of nature on ruins—not only accentuates the romantic value of the ruin but is also an indicator of the degree of decadence of the civilizations that once built these same structures. Thus, as Pramod K. Nayar claims for the case of India, while the European ruin "symbolized antiquity, continuity, and a venerable tradition, the Indian ruin seemed to symbolize a decadent social order" (65). In the latter, but not in the former, the conjunction of flora, fauna, and ruins came to be analogous with decay, disorder, and negligence.

In this regard, Hiram Bingham's description of his first encounter with Machu Picchu in 1911 offers a fascinating outline of the embrace of nature in the discursive construction of this site as a lost city. In describing his first impressions of the Machu Picchu ruins, Bingham writes:

> I entered the untouched forest beyond, and suddenly found myself in a maze of beautiful granite houses! They were covered with trees and moss and the growth of centuries, but in the dense shadow, hiding in bamboo thickets and tangled vines, could be seen, here and there, walls of white granite ashlars most carefully cut and exquisitely fitted together. (*Inca Land,* 320)

Previously in the book, Bingham has offered some information about the settlers—mostly Indigenous peasants—that live in and around what is nowadays known as the Machu Picchu area. At the same time, he reports that several terraces have been "recently rescued from the jungle," and that a section of the forest composed of large trees has been chopped and burned down "to make a clearing for agricultural purposes" (320). And yet, at least for Bingham, Machu Picchu is not part of the area where current settlers have planted maize, sweet and white potatoes, sugarcane, beans, peppers, tree tomatoes, and gooseberries. Instead, Machu Picchu ruins are in what has been left "untouched" (319), in that zone at the edge of the cultivated parcels, that has been covered almost entirely by bamboo thickets and tangled vines.

Of course, it is not just that this vague allusion to wild vegetation contributes to enhancing the romantic or even picturesque experience of Machu Picchu. More importantly, the emphasis that Bingham puts on this wild or semi-wild forest that surrounds the site is what justifies the treatment of Machu Picchu as a ruin. Ultimately, this is a rhetorical matter. By using vegetation as part of his description, Bingham reproduces one of the many codes that are part of the modern iconography associated with understanding ruins in postcolonial scenarios. Therefore, beyond whether the vegetation described by him existed or not, what is interesting to consider is the way in which the mention of these overgrown structures helps to accentuate the image of decay of the site. From this perspective, Machu Picchu can be interpreted as a ruin because it coincides with modern expectations of how ruins of decaying or declining civilizations should look, and, for all that matters, the references to the spontaneous growth of vines, bushes, and thickets ensure this effect.

It is possible to gain a better understanding of the importance that this element acquires within Bingham's work when we consider the available versions of his first encounter with Machu Picchu. Along with other significant changes that the description of this event underwent over several decades, it is quite remarkable how Bingham gradually modulated the amount of vegetation present in the area to make the impact of natural forces on the site increasingly evident, and with it, its degree of decay. Thus, in his early article, "Vitcos: The Last Inca Capital," published in 1912, he simply states: "When I first saw the ruins of Machu Pichu, which is on a very high mountain commanding a magnificent view . . . I thought that I must have come across Pitcos" (174). And nothing else. However, in the version of 1913, published by the *National Geographic Magazine*—and for which he received help from Gilbert Grosvenor, director and editor of the magazine—the description of this same event includes, for the very first time, an allusion to the vegetation and, above all, a characterization of the forest that surrounds the ruins in the terms of "a tropical forest, beneath the shade of whose trees we could make out a maze of ancient walls" (*In the Wonderland*, 408). The evolution of this tendency finds a perfect closure in his most famous work, *Lost City of Incas* (1943):

> Suddenly I found myself confronted with the walls of ruined houses built of the finest quality of Inca stone work. It was hard to see them for they were partly covered with trees and moss, the growth of centuries, but in the dense shadow, hiding in bamboo thickets and tangled vines, appeared here and there walls of white granite ashlars carefully cut and exquisitely fitted together. (152)

The shift from "they were covered" (320) to "hard to see them" is not a minor modification. This seals the image of ruins covered almost entirely by "the growth of centuries" that was barely insinuated in prior versions. Furthermore, moss is incorporated in this last version as part of the delicate balance of natural elements that have come to occupy the empty place left by humanity after several centuries of neglect and oblivion. The resulting scenario, however, is not a natural garden—as in many representations of English ruins. By contrast, for Bingham, these ancient buildings were literally devoured by the "jungle," a term that has a negative connotation and refers to "disorder and disorienting growth" (Slater 117). Constructed in this way, the scenario is reduced to an empty space, a kind of desert where nature has come to rule again. This reversion to a primordial wilderness is what gives its ruined aura to Machu Picchu. And, under this light, Machu Picchu is nothing but a set of ancient structures covered by an untamed and chaotic nature.

John Beasley-Murray has pointed out that Bingham's real achievement was "less to *discover* Machu Picchu than to put it into discourse: to articulate its stones, to make them speak in the recognizably modern idiom of ruination" (218). However, to obtain this effect, it was also essential for him to organize the entire scene within a framework of vision that would allow the site to be understood as a ruin. To this end, the inclusion of explicit references to the wild and overgrown vegetation was crucial, despite the fact that Indigenous peasants continued to use the area for housing and agriculture. Machu Picchu, at least the Machu Picchu that Bingham created, is indeed a lost city: ruins of what was once a vibrant and cultural hub buried under layers and layers of vegetation.

Rui(nation) and "the Embrace of Nature"

Bingham's approach to Andean ruins is not a unique or exceptional undertaking. In fact, in 1912, the same year that Bingham embarked on his second trip to the area with the Peruvian Yale Expedition, Peruvian writer José de la Riva-Agüero also traveled through the highlands of southern Peru. Unlike Bingham, whose central objective was the search for "lost cities," Riva-Agüero's journey was a pilgrimage to the roots of the nation, which he believed were a combination of both Spanish and Inca imperial legacies as essential components of what he called the "Integral Peruvian identity" (Vich). Continuing a long tradition of travelers and naturalists who combined the discourses of geography, history, and literary representation of nature, Riva-Agüero conducted a general diagnosis of the state of *La Sierra* where both nature and ruins are interpreted as indicators of Peru's failure to consolidate the project of becoming a nation based on integral Peruvian identity. Perhaps there is no better example of how these elements are articulated within his concern for the future of Peru than his description of the ruins of Vilcashuamán:

> The afternoon was advancing; and I took advantage of the remains of the day to see the ruins of the watchtower and the palace.... Through the alley, the pigs were rooting through the mud in front of the huts. A couple of seats had fallen at the foot of the tower.... Among the cracked stones some plants had grown, and their leaves fluttered in the open air. (99)

As Riva-Agüero informs us, Vilcashuamán or Vilcas was, during the height of the Inca Empire, an important administrative and cultural center due to its strategic location at the intersection of the trade routes that crisscrossed the Inca Empire. Parades, rituals, opulent celebrations, and other demonstrations of power and richness were frequent in the city. But not anymore. By the time Riva-Agüero visits this site, Vilcas is barely a shadow of the past, a town full of ruins with a long history of wars, oblivions, and silences on its shoulders. Vilcas, however, is not an abandoned town, at least not in the strict sense of the term. Villagers—mostly *morochucos,* shepherds of the plains of Peruvian Andes—have built their houses around and over the old buildings. In many cases, new houses were built with pieces and fragments taken from the old buildings, making it almost impossible to distinguish where the ruins begin and where the town does. As John Beasley-Murray notes ironically, "One might also say that the place is a set of ruins that enclose a town, as it can be hard to judge where the ruins end and the town starts" (217). In a way, Vilcas is made of recycled materials; Vilcas is the past recycled.

Several reasons contributed to the decline of this imperial city. For Riva-Agüero, the most important dates back to the colonial period. With the Spanish domination, specifically with the foundation of the city of Huamanga, Vilcas began to experience an economic downturn, which would reach its peak during the early years of the Republic. Nature, therefore, does not seem to have played a relevant role in this case. And yet nature appears in *Paisajes peruanos* in a less explicit but significant way, functioning as a powerful indicator of Vilcas' ruinous fate. Víctor Vich, among many other critics, has underscored that the depiction of Indigenous figures in *Paisajes peruanos* lacks depth, reducing them to mere decorative elements within the natural landscape of *La Sierra*. Implicitly, but repeatedly, Riva-Agüero assigns to them a status similar to that of trees, stones, or nonhuman animals. Just like the moss in Bingham's portrayal, morochucos are a natural entity, primarily decorative, yet possessing the quasi-agential function of destroying and degrading the built environment.

None of this applies, however, to the Inca ancestors of these communities whom Riva-Agüero considers paradigmatic examples of civilization and culture. In fact, Riva-Agüero conceives current settlers as a degraded version of his admired Incas, where this degradation is analogous to a sort of feralization, a regression to a nature-like way of living. This vision becomes particularly evident when Riva-Agüero addresses the issue of the hygiene of the morochucos. On this matter, he comments that, as in the fall from ancient Rome to the European Dark Ages, "degradation should have caused them to lose the habits of cleanliness" (99). But the lack of hygiene is just a symptom of a major, and more extended, disease. Morochucos are a corrupted community in which time has erased any trace of civilization or culture. In sum, they are much like ruins: remains of an empire that was once glorious, now reduced to mere nature.

Thus characterized, it is not surprising that the morochucos had played an active role in the ruinous state of Vilcas: "The inhabitants, neighbors, and others related to the morochucos (famous for their robberies and bravery) have finished ruining the old buildings by removing the stones to use them in the construction of their homes and courtyards" (98). Like the wild vegetation of Machu Picchu, the presence of the morochucos is, in this case, a sign of decay. They are mere forces that erode the masonry in much the same way that weeds crack the cobblestones and moss deteriorates the facade of an important sanctuary. To use a contemporary term, they are a biocolonizer species—that is, a foreign biological community, like moss, lichen, or fungi, with biodeterioration abilities to cause an undesirable change in the properties of a material (Pinna and Salvadori 15). To some extent, morochucos have become accomplices of nature or, which means the same thing in this case, enemies of civilization.

Víctor Vich (127) observes that *Paisajes peruanos* aestheticizes Peruvian history by reducing its inherently conflictive nature to the most acceptable form of nostalgia for imperial times. The case of Vilcas is a clear example of this. Reduced to ruins, the town becomes a repository of longings for the past, a "poignant symbol of abandonment and dejection" (Riva-Agüero 99). However, this effect is the result of the interweaving of natural elements (mainly morochucos) as part of the description of the state of the town and its people. Through this, Vilcas can be associated with decadence, abandonment, or simply degradation, ultimately becoming a ruin: "I have never felt a more piercing and heartbreaking sensation of decay" (99), declares Riva-Agüero while observing the imperial plaza of Vilcashuamán, now invaded by birds, chickens, herds, and of course morochucos.

Both approaches are an expression of the modern mentality for which nature and culture are two different and mutually exclusive categories. One specific dimension that this view takes coincides with what Neil Smith calls the "externalization of nature"—that is, the idea that nature corresponds to a "realm of extra human objects and processes existing outside society. External nature is pristine, God-given, autonomous; it is the raw material from which society is built, the frontier which industrial capitalism continually pushes back" (11). In addition to its various effects on different cultural fields, this view of nature has been crucial for the modern perception of ruins. In his famous essay "The Ruin" (1965 [1911]), George Simmel points out that ruins represent a reversal of the cosmic order where the spirit (*Geist*) dominates nature. This reversal allows "the brute, downward-dragging, corroding, crumbling power of nature" (261) to shape what was originally built by human hands. In this context, the fascination that ruins evoke—what I previously described as the Ozymandias effect—is essentially a memento mori: a reminder that all works of the spirit will eventually return to nature.

It is notable that for Simmel, ruins—European ruins—reveal a tension between the forces of nature and the spirit, as if something of the original spirit always remains in each ruin overtaken by nature. However, this perception changes substantially in colonial and postcolonial contexts. In these scenarios, descriptions of buildings reclaimed by vegetation or inhabited by wild animals are symptomatic of the indisputable triumph of nature over the will of the historical inhabitants of these sites. As a result, a void is created, effectively erasing any trace of humanity because, if nature and culture are two irreconcilable opposites, the presence of one will logically imply the absence of the other.

Needless to say, this is precisely what happens in the cases of Machu Picchu and Vilcashuamán. Bingham and Riva-Agüero's depictions of ancient buildings overtaken by vegetation, animals freely roaming the squares, and Indigenous shepherds using ancient monuments as their homes all serve to imply—given that these elements are categorized as natural forces—the absence of civilization and, at times, even humanity from these sites. If ruins are what remain after the clash between nature and culture, then Peruvian ruins are the undeniable evidence of the disappearance of civilization from this land.

Postcatastrophic Visions of *La Sierra*

There is an additional dimension that needs to be considered here, since neither Bingham nor Riva-Agüero speak exclusively of archaeological ruins, but rather of the Peruvian Sierra as a landscape in ruins. Throughout *Inca Land* and *Paisajes peruanos,* readers are exposed to a parade of vestiges, loose fragments, rubble, and remains of what seems to be a larger and now destroyed totality. These scattered fragments give the impression that *La Sierra* is a decadent and miserable landscape.

Riva-Agüero is probably more explicit than Bingham in this respect. In a letter to his friend José Gabriel Cossio, the "general impression" of the landscape emerges with extraordinary clarity:

> The general aspect of this part of the territory and its cities is desolate, because of how broken and rugged, because of the narrowness of the valleys, the unusable nature of the rivers that run so deep that they cannot be used for irrigation, and because of its poor and decadent populations. . . . Ruined houses, closed churches, abandoned convents; everything gives the impression of a Spanish past irretrievably lost and degraded. . . . Peru, my friend, is a ruin, a country of memories. (*Epistolario,* 1103–4)

Riva-Agüero's frustration can largely be attributed to his unwavering belief that human beings should have complete control over nature for their own benefit and financial gain. Throughout his work, he depicts nature as a force that acts independently, seemingly unconcerned with human needs or desires. Furthermore, he frequently mentions natural disasters such as droughts, floods, and landslides as proof that the area is governed by natural forces beyond human control. As a result, the landscape is portrayed as a desolate wasteland, where civilization has retreated, and nature is left to flourish aimlessly.

There is a similar line of thought in Hiram Bingham. When evaluating the current conditions of *La Sierra*, Bingham writes:

> Food is hard to get. Few crops can be grown at 12,500 feet. Some barley is raised, but the soil is lacking in nitrogen.... Naturally these Indians always feel themselves at the mercy of the elements. Either a long rainy season or a drought may cause acute hunger and extreme suffering. (*Inca Land* 103)

Like Riva-Agüero, Bingham draws a sharp contrast between the current inhabitants of the Sierra and the Incas, whom Bingham describes as a civilization "marked by inventive genius, artistic ability, and knowledge of agriculture that has never been surpassed" (*Lost City of the Incas*, v). However, in this case, the difference is structured around a corpus of specific knowledge about the Andes and its ecological particularities, which finds expression in what John Murra popularized through the concept of "vertical control of ecological zones" (60–61). Adopting the conventional definition that posits pre-Columbian societies established colonies in distant and often noncontiguous ecological zones to access goods produced in those areas, what sets apart the present and past in *La Sierra* is precisely their adeptness at implementing such vertical control. This, of course, yields a stark contrast between *La Sierra*'s past and the present landscape. While the landscape of imperial times is a highly controlled and managed environment characterized by neat and orderly fields, well-maintained roads and buildings, and carefully managed waterways, the current landscape is presented as a degraded, semiwild landscape where the neglect and ineptitude of the inhabitants have allowed nature to invade and reduce the great works of the past to ruins. In this scenario, the landscape is largely overrun by invasive vegetation, and the fields and waterways have become choked with debris.

An essential aspect of Riva-Agüero's and Bingham's narratives is that they present the abject state of *La Sierra* as if it were the result of a catastrophe, a colossal disaster, or, as Riva-Agüero himself expresses it, "a historical shipwreck" (100). In this sense, their descriptions are postcatastrophic representations of *La Sierra*—that is, representations that deal with the ruins and aftermath left by an event conceived as "eschaton, as an end of something, a way of life or thinking" (Berger 5). There is some consensus that an event of this magnitude in the case of *La Sierra* was the Conquest (and, with it, colonialism) since it involved the extermination of millions of Indigenous people as well as the destruction of their societies and cultures (Quijano 13). Given the importance and magnitude of this event, it is surprising that references to it are rather scarce and almost always marginal in Riva-Agüero and Bingham's accounts. This omission generates a kind of interruption in the historical continuum in which there seems to be no relationship between the world of the Incas and the world of present-day peasants, no causal sequence that explains the passage from one state to the other. Often, this can be explained in the form of a "cognitive dissonance" that "breaks with conventional understandings of causality and temporal sequence" (Beasley-Murray 224). But what is interesting is that this historical ellipsis creates the impression that the catastrophe is fully explained by its effects—that is, that the decline of Andean societies is essentially a matter of their inability or lack of knowledge to dominate natural forces.

In her book *Monsters, Catastrophe, and the Anthropocene*, Gaia Giuliani introduces a provoking distinction that can shed some light on the case of *La Sierra*. Drawing on a postcolonial analysis of how modernity imagines catastrophes and monstrosity, Giuliani differentiates between "places of disaster" (civilized spaces where disasters are considered isolated and containable events) and "places for disaster" (postcolonial spaces where disasters are seen as inherent and permanent). Certainly, *La Sierra*, as portrayed by Bingham and Riva-Agüero, is a place of disaster. However, Giuliani's classification reveals something else. In particular, it shows how modernity evades its share of responsibility by attributing the precarious state of peripheral regions to a lack of modernity. Therefore, it is not modernity, with its dark side expressed in colonialism, that produces the postcatastrophic landscape of *La Sierra*. Instead, it is precisely the absence of modernity that explains the catastrophe, its overgrown ruins, the expansive nature, and its widespread decay.

Viewed from this perspective, *La Sierra* functions as an inverted mirror: through it, it is possible to see one of the many ways in which modernity envisions its transcendence, its own end, and with it, what a world without it might look like. In a way, this is one of the many strategies by which narratives of modernity distance themselves, both spatially and temporally, from what is deemed "not modern," thus erasing and denying their own genealogy and ongoing colonial dark side. In light of this, it could be said that the postcatastrophic Sierra of Riva-Agüero and Bingham reveals itself at the same time as trauma and sublimation for the modern gaze. It is trauma because it expresses modern fear of the end of civilization understood as a loss of control over nature, but it is also sublimation because this fear is somehow contained in a postcolonial scenario without sufficient technology, knowledge, or ability to dominate and control nature.

Ozymandias

In his work *After the End: Representations of Post-Apocalypse,* James Berger contends that "the most dystopic visions of science fiction can do no more than replicate the actual historical catastrophes of the twentieth century" (XIII). Although this may be the underlying premise of the current chapter, my own idea extends beyond Berger's proposed timeline. I firmly believe that dystopian images of the future are also influenced by descriptions of catastrophic events or apocalypses that date back even further in time. *La Sierra,* as discursively constructed by Riva-Agüero and Bingham, provides a fitting example of this phenomenon. Therefore, in this final section, I would like to highlight certain points where the postcatastrophic vision of *La Sierra* presented by both authors converges with the portrayal of a future world composed of overgrown ruins—a description that, as I mentioned at the outset, encapsulates some of the recent concerns regarding the future of the planet.

Perhaps the most obvious connection between both visions is that they are structured around some form of externalization of nature and, therefore, embody the modern division between nature and culture. This, of course, confirms the idea that contemporary narratives on environmental catastrophes are quintessentially modern (Bettini 39), but also suggests that behind the visions of weed-grown cities, the old antagonisms of modernity still linger. Thus, akin to *La Sierra,* the fundamental fear here is not so much that flourishing vegetation invades the spaces of civilization, but, rather, what this disruption indicates—namely, the collapse of civilization.

The specific case of *La Sierra,* however, adds an extra layer to the discussion. Although modernity brought about a division between places of disaster and places for disaster, with the Anthropocene this distribution is no longer sustained, producing a sort of planetarization of disaster and catastrophe (Giuliani 159). I believe that it is precisely this fact that encourages the constant and latent prospect of a near future where nature overwhelms not only peripheral societies but also the centers of civilization. If so, the fascination and rejection that buildings covered by lush vegetation elicit in our time is nothing more than an expression of the anxieties that arise from the possibility of becoming the object denied and repressed by the modern history of the West. Of course, I am referring to colonialism and its ruins.

Note

1 Geographically situated between the Coast and La Montaña, La Sierra is one of the three primary regions into which the Peruvian territory has been geopolitically divided. Allegedly, what sets it apart from the other two is, on the one hand, its high-mountain landscape, and on the other, the enduring presence of Indigenous communities with cultural and social roots dating back to pre-Columbian times.

Works Cited

Beasley-Murray, John. "Vilcashuamán: Telling Stories in Ruins." *Ruins of Modernity*, edited by Julia Hell and Andreas Schönle, Duke UP, 2010, pp. 212–31.

Bettini, Giovanni. "Environmental Catastrophe." *Companion to Environmental Studies*, edited by Noel Castree, Mike Hulme, and James D. Proctor, Routledge, 2018, pp. 29–42.

Berger, James. *After the End: Representations of Post-Apocalypse*. U of Minnesota P, 1999.

Bingham, Hiram. *Inca Land; Explorations in the Highlands of Peru*. Houghton Mifflin, 1922.

Bingham, Hiram. "In the Wonderland of Peru—The Work Accomplished by the Peruvian Expedition of 1912, Under the Auspices of Yale University and the National Geographic Society." *National Geographic Magazine*, vol. 24, 1913, pp. 387–573.

Bingham, Hiram. *Lost City of the Incas: The Story of Machu Picchu and Its Builders*. 1943. Atheneum, 1963.

Bingham, Hiram. "Vitcos: The Last Inca Capital." *American Antiquarian Society*, vol. 22, 1912, pp. 135–96.

Castro-Klarén, Sara. "The Nation in Ruins: Archaeology and the Rise of the Nation." In *Beyond Imagined Communities: Reading and Writing the Nation in Nineteenth-Century Latin America*, edited by Sara Castro-Klarén and John Charles Chasteen. Johns Hopkins UP, 2004.

Dwyer, John. "Ruins and the Embrace of Nature." *Australian Garden History*, vol. 21, no. 2, 2009, pp.10–14.

Fisher, Mark. "What Is Hauntology?" *Film Quarterly*, vol. 66, no. 1 (Fall 2012), pp. 16–24.

Flores Galindo, Alberto. *Buscando un Inca: Identidad y utopía en los Andes*. Casa de las Américas, 1986.
Giuliani, Gaia. *Monsters, Catastrophes, and the Anthropocene; a Postcolonial Critique*. Routledge, 2021.
Heffernan, Teresa. *Post-Apocalyptic Culture: Modernism, Postmodernism, and the Twentieth-Century Novel*. U of Toronto P, 2008.
Horn, Eva. *The Future as Catastrophe: Imagining Disaster in the Modern Age*. Columbia UP, 2018.
Jonk [Jonathan Jimenez]. *Naturalia: Reclaimed by Nature*. Carpet Bombing Culture, 2018.
Kirschbacher, Felix. "More than Ruins: (Post-)Apocalyptic Places in Media." In *Ruin Porn and the Obsession with Decay*, edited by Siobhan Lyons, Palgrave Macmillan, 2018, pp. 201–16.
López Lenci, Yazmín. *El Cusco, paqarina moderna: Cartografía de una modernidad e identidades en los Andes peruanos (1900–35)*. Fondo Editorial Universidad Nacional Mayor de San Marcos / CONCYTEC, 2004.
Murra, John. *Formaciones económicas y políticas del mundo andino*. IEP Ediciones, 1975.
Nayar, Pramod K. *Colonial Voices: The Discourses of Empire*. Wiley-Blackwell, 2012.
Painter, James. *Climate Change in the Media: Reporting Risk and Uncertainty*. Reuters Institute for the Study of Journalism, 2013.
Pinna, Daniela, and Ornella Salvadori. "Processes of Biodeterioration: General Mechanisms." In *Plant Biology for Cultural Heritage: Biodeterioration and Conservation*, edited by Giuliua Caneva et al. Getty Conservation Institute, 2008, pp. 15–34.
Quijano, Aníbal. "Colonialidad y modernidad/racionalidad." *Perú Indigenista*, vol. 13, no. 29, 1992, pp. 11–20.
Riva-Agüero, José de la. "Carta a José Gabriel Cossio." *Epistolario*, vol. 2. Pontificia Universidad Católica del Perú, 1962, pp. 1103–1104.
Riva-Agüero, José de la. *Historia del Perú*. Vol. 2. Librería Studium, 1953.
Riva-Agüero, José de la. *Paisajes peruanos*. Pontificia Universidad Católica del Perú, 1969.
Simmel, Georg. "The Ruin." *Essays on Sociology, Philosophy and Aesthetics*, Edited by Kurt H. Wolff. Harper & Row, 1965, pp. 259–66.
Slater, Candace. "Amazonia as an Edenic Narrative." *Uncommon Ground: Rethinking the Human Place in Nature*, edited by William Cronon, W. W. Norton, 1996, pp. 114–31.
Smith, Neil., et al. *Uneven Development: Nature, Capital, and the Production of Space*. U of Georgia P, 2008.
Tsing, Anna Lowenhaupt, et al. "Introduction: Haunted Landscapes." *Arts of Living on a Damaged Planet*, edited by Anna L. Tsing et al. U of Minnesota P, 2017, pp. 1–14.
Vich, Víctor. "Vicisitudes trágicas: territorio, identidad y nación en los *Paisajes peruanos* de José de la Riva-Agüero y Osma." *Revista Andina*, no. 32, 2002, pp. 123–34.

2

The Political Ecology of Artistic Forms

Culture, Extractivism, and Environmental Imagination

4

What Is the Conflict About?

Ecology, Ontology, and Decoloniality in
Joseph Zárate's *Guerras del Interior*

José Carlos Díaz Zanelli

In recent years, the environmental humanities have evolved toward a productive disciplinary entanglement that allows scholars to approach contemporary ecological and environmental concerns with theoretical flexibility and critical openness.[1] One of the most appealing aspects of this cross-disciplinary field is the possibility of thinking about literary representations of nature, a foundational purpose of ecocriticism, in tandem with the representation of Indigenous mobilizations and, consequently, the anthropological reformulations to which are forced non-Indigenous scholars who aspire to a proper understanding of the Indigenous imaginary in a context of ecological crisis. For this reason, this work considers it particularly pertinent to bring to the table of literary criticism a theoretical perspective that cultural anthropologists have labeled in recent years as the "ontological turn," a merging that will be elaborated throughout this work.

This chapter illuminates what is represented when a cultural text portrays conflicts that engage Indigenous peoples and their natural settings when extractive industries threaten the latter. The Latin American media and academia have labeled these types of confrontations as *socioenvironmental conflicts*, a neologism that semantically challenges the separation between society and nature as an ideological binarism that structures the idea of development imposed over the socionatural settings of Indigenous communities. This imposition causes the conflicts that cultural anthropology examines through the lens of ontology (De la Cadena; Blaser "Ontological"), exploring the realms of Indigenous political philosophy and foregrounding the political praxis of affected communities. Hence, one of the first aspects we must recognize in these conflicts is that there is no consensus about what is in dispute. Then what are these conflicts about? In what follows, I answer this question in three specific ways—ecologically, ontologically, and decolonially—by navigating through the book chronicles *Guerras del Interior* (2018) by Peruvian writer and journalist Joseph Zárate.[2]

This book portrays three peasant and Indigenous activists from the Peruvian Andes and Amazon. It is divided into three parts, each titled with an element that articulates the extractive industries involved in each case: wood, gold, and oil. The first chapter ("Wood") presents the story of Edwin Chota, an Indigenous ecological activist and a member of the Asháninka nation whose territories span from the central Peruvian rainforest to the Brazilian one. Chota was the leader of the Alto Tamaya-Saweto native community and fought against illegal logging mafias by organizing Indigenous communities to defend the rainforest. Amid this struggle, he was murdered by these mafias along with three other Indigenous activists with whom he was demanding the titling of Asháninka's land.

The second chapter ("Gold") recounts the case of Máxima Acuña, an Andean peasant farmer who owns farmland in the rural area of the department of Cajamarca and whose land has been threatened in recent years by the expansionist interest of the Conga Mine, a project that jeopardizes Acuña's properties and the Conga Lake, which would disappear with the mine expansion. This project is funded by American and Peruvian mining corporations, a coalition of companies backed by the Peruvian state. The history of conflicts between Acuña and the Conga Mine includes a set of violent episodes such as controversial judicial resolutions in favor of the company, the destruction of Acuña's house, the killing of her cattle, and physical attacks by the police and the mine's staff against her and her family. At the moment, the project is on hold and under legal dispute.

The third chapter ("Oil") narrates the story of Osmán Cuñachí, an Awajún child from the Nazareth community who, in 2016, was recruited by the staff of the state-owned petroleum company Petroperú, along with other Indigenous individuals, to stop a leak of more than five hundred thousand liters of oil in the middle of the Amazon rainforest. The oil spill originated in the Norperuano Oil Pipeline that crosses the country from west to east, transporting oil extracted from the Amazon toward the refineries of the coast. Through this child's story and the health problems caused by his direct exposure to crude, Zárate showcases the environmental impact of oil drilling on the Amazonian Indigenous communities.

In the following sections, I analyze specific instances of the conflicts reflected in *Guerras* from an ontologizing perspective, focusing mainly on Acuña's and Chota's Indigenous ecological activisms. Mario Blaser ("Reflexiones") argues that in socioenvironmental conflicts involving Indigenous peoples, where the institutions of modernity (such as the state, corporations, and legal systems) view nature through a framework of rational understanding—what he calls "rational politics"—Indigenous political modalities are excluded from the discussion.[3] Such exclusion of Indigenous rationalities generates a lack of consensus regarding what is at stake because the Indigenous way of conceiving nature is silenced. Therefore, the conflict can no longer be explained purely in political or cultural terms but on the ontological ground, since the parties in conflict do not agree on the existence of what is disputed (70).[4] I would add that this type of conflict occurs even in those cases where Indigenous activists are seemingly involved in the negotiations before the physical confrontation, since they are included as human entities but not their Indigenous ontologies, which, as we will see later, may engage with different senses of humanness. In such cases, Indigenous activists fulfill the role of what Jorge Marcone calls "stone guests" ("The Stone" 228) because, although they have a seat at the decision-making table, Indigenous ontologies are not validated in this process, since they are typically judged as fixed and premodern worldviews.

The Limits of Culturalist Explanations

In *Wars of the Interior*, the stylistic characteristics of Zárate's chronicles fluctuate between narrative fragments and informative passages influenced by his journalistic background:

> In the Andean world, there is a long tradition of land transfer under peasant norms. . . . A peasant community is defined as an organisation of families who inhabit and control land, and who are linked by ancestral, social, economic and cultural ties. The members of the community have no individual property deeds: the land belongs to all of them. . . . "Modern Peru," however, is governed from the cities, and land is bought and sold independently of how it's used, worked, and inhabited. Land is a possession, just another product. (*Wars* 99–100, my ellipses)

Occasionally this narrative sequence is interrupted by aphorisms with which the author rounds up some of the ideas outlined in the narrative excerpts. This style is illustrated in the above quotation, in which Zárate contextualizes the notion of inseparability between peasant communities and their land to explain the motivations for Acuña's struggle. This particular inseparability clarifies that the notion of land as a commodity does not exist in peasant communities because, to them, the land is connected to collective belonging—that is, to the existence of individuals as inseparable elements of a collectivity composed by humans and the landscape where they work, which converts land into something more than a lifeless object. Subsequently, we can infer that Indigenous communities do not commodify the components derived from land that modern economic logic refers to as "raw materials." The land defines the identity of the *comuneros;* without it they are no longer comuneros. In contrast, in modern Peru, the commercialization of land does not alter the social identity or collective belonging of individuals. Cultural relativism may contend that the differential variable is a matter of identification with the land. However, as will be developed later, the land is a constitutive component of beingness construction rather than a mere item of external identification.

The aforementioned demarcation reveals the limitations of the culturalist discourse to explain the elements involved in the struggle of Indigenous activists, not only Andeans like Acuña but also Amazonians like Chota, when explaining that for the Asháninkas "no one owns the land or the hunting or fishing sites. The idea of individual private property is alien to them" (*Wars* 25–26). The divergence between the territorial sense of these Indigenous activists and modern Peru cannot be explained only by social codes commonly defined as culture. The idea of culture as a dividing category that explains the differences in behavior between different social groups has been deeply disputed by recent developments in cultural anthropology.

In her elaboration on Indigenous cosmopolitics, Marisol de la Cadena warns about the conceptual challenge of understanding "sentient entities" (e.g., landscapes, mountains, lakes, trees—Latour's *actants*) as material-spiritual beings threatened by capital and the neoliberal state.[5] She proposes that the nonhuman entities involved in Indigenous political activisms "are contentious because their presence in politics disavows the separation between 'Nature' and 'Humanity,' on which the political theory our world abides by was historically funded" (342). Thus, the Andean and Asháninka liaisons with nature embodied by Acuña and Chota represent more than cultural identity politics. This work argues that the political mobilization of these two Indigenous ecological activists is fostered by the ontological commitment to their socionatural settings that is framed in a long tradition of Indigenous decolonial struggle. Moreover, their activism is not inspired by the loss of a natural resource in a conventional materialist sense, nor by a sudden environmental consciousness, but by the imminent risk that extractive industries provoke against the world-making praxes that ground their beingness. Extractivism threatens what could be defined as their "ontological identities," which this work elaborates on in the next section.

The transfer of land described by Zárate under Andean peasant norms is not a commercial transaction, but, rather, the germination of human entities within a sense of belonging that transcends the conventional limits of cultural politics due to the inevitable engagement with nonhuman actants. However, it is not my intention to assert that the agency of these actants depends exclusively on their interaction with human beings. As Métis anthropologist Zoe Todd (245–46) argues, the locus of agency is crucial to understanding the existential independence of the elements framed within the Anthropocene by the ontological turn. Yet, in this specific work, my interest focuses on this interaction as it contributes to reevaluating the analytical imperatives with which socioenvironmental conflicts are interpreted, paying attention in this case to Acuña and Chota. Blaser ("Reflexiones" 70) notices that categorizing nonhuman entities as culture can make us fall into the modern trap of a new form of cultural relativism. He explains that reading ecological conflicts with the participation of nonhuman entities as discrepancies where there is a unique idea about nature would lead us to read these conflicts as epistemological disputes, omitting the lack of consensus inherent to ecological conflicts. Zárate spends much of the chronicles about Acuña and Chota to illustrate the degree of connection that the former has with her crops, lakes, rivers, fountains, and animals, and the latter with his forest and its functionality in defining the Asháninka identity. In *Guerras*, neither of these two activists is mobilized by an environmentalist discourse in which the only discrepancy with modern Peru would be what to do with nature; ultimately, there is no consensus about what nature is. However, for Acuña and Chota, the elements that for modern rationality dwell in a distant natural wilderness in the Amerindian context are nonhuman entities with an agency within their respective Andean and Amazonian sociabilities, where these nonhuman entities fulfill a crucial role in the definition of their identities and their belonging to collectivity.

The Blast of Ontologies

The hostile atmosphere created by the advance of extractivism and deforestation in the Amazon rainforest is understood through Asháninka cosmology as a form of political rivalry that violently encroaches upon the realm of ontology. For instance:

> The Asháninka have a profound belief in the power of evil. From them, Edwin Chota learnt that here were invisible enemies in the world and that he had to defeat them as well. The elders call them "kamári": demons. Spirits that hide in the forest, in the caves. Malevolent beings that grind people's bones, that suck out their eyes. They can kill a newborn baby or the strongest warrior. They can possess a person, Asháninka or not, and make them murder their own brother without a second's thought. Kamári are the essence of evil, and the illegal loggers, like the terrorists before them, are some of their more recent incarnations. (*Wars* 28)

The imaginary of evil embodied in the kamári and the associations that the Asháninka people make of these spirits with the outsider entities that endangered their existence (e.g., the terrorist group Shining Path in the 1980s, and the timber logging mafias in the 2000s) highlights, on the one hand, the relationality that structures the Asháninka cosmology. Fernando Santos-Granero (273–75) has extensively elaborated on the role of the kamári in the Asháninka moral code and belief system, portraying them as demonic teachers who morally corrupt children by initiating them into witchcraft or dehumanizing them. But, on the other hand, in the light of Asháninka political thought and Chota's ecological activism, this fragment also exhibits what Juan Ricardo Aparicio and Mario Blaser call "the insurrection of subjugated knowledge," which they define as patterns of insurrection promoted by Indigenous cosmologies that colonial differences label as "unrealistic," but that actually embody "worlds and knowledge otherwise" (70). Thus, we can read Chota's ecological activism as a form of Indigenous mobilization that takes place not only in another socionatural reality but also in its world-making process. This Indigenous socionatural reality brings ontological commitments that diverge from the world-principles of modernity/coloniality, decolonially speaking. Since culture as a category of analysis is insufficient to explain these divergences, the conflict embodied between the Asháninka cosmologies and the threat of the various kamári, reveals a blast of ontologies—that is, a conflict between oppositional world-making modalities, which I propose to examine through the lens of the anthropology of ontologies.

Analyzing a corpus of documentaries that focus on the tangible struggle of popular environmentalism, Marcone criticizes that these films represent Indigenous activists as "just limited illustrations of larger more-than-environmental social movements whose existence and political ontologies, paradoxically, remain silenced in the films although they underlie them" ("Filming" 209). By answering the question of what the conflict is about, this work precisely illuminates in *Guerras* those political ontologies that these environmentalist films overlook. This is the reason why I consider it necessary to engage this type of analysis with the ontological turn. Eduardo Kohn defines the anthropology of ontologies as a nonreductionist ethnographic study that analyzes realities not circumscribed to human worlds. For him, in this field, ontology refers to the study of different realities, and, hence, it is not constrained to the study of the world and humans but goes further without leaving humans behind. Ultimately, "it is about what we learn about the world and the human through the way in which humans engage with the world. Attention to such engagements often undoes any bounded notion of what the human is" (313). Tacitly, Kohn's definition of the anthropology of ontologies connects to the idea of personhood formulated by Eduardo Viveiros de Castro's Amerindian perspectivism ("Exchanging"; "Cosmological") in which the different entities that dwell in the world (animals, humans, spirits) are perceived among members of the same species as persons—that is, as conferring a degree of humanity to each other, but with special attention being paid to the world-making process. Through participatory research in South America, Kohn and Viveiros de Castro have arrived at these definitions after anthropological fieldwork in dialogue with local Indigenous communities. This approach enables them to reframe the analytical categories that organize Indigenous studies, elevating them beyond mere storytelling to ethnographic value.

The political interrelations between these different forms of humanness constitute an Indigenous cosmopolitical assembly in which human and nonhuman entities forge an alliance of solidarity in the struggle against a common enemy: extractivism. In what follows, I will refer to this concept by borrowing the definition of Eduardo Gudynas, who refers to it as the extraction of natural resources that generates a transformative impact on nature and, usually, involves a degree of violence and rights violations.

To re-create Chota's portrait, Zárate interviewed Francisco Berrospi, the only environmental prosecutor assigned to Ucayali (a region of the Peruvian Amazon) who paid attention to Chota's demands. Recounting his interactions with the Indigenous activist, Berrospi says: "Edwin had a very strong connection to the forest.... And he knew how to get it across" (*Wars* 32). With the help of Berrospi's testimony, Zárate reconstructs a scene in which these two men were in Ucayali, walking through a lumber mill, and Chota asked Berrospi to touch a huge trunk of *shihuahuaco* (a species of tree native to the Peruvian, Brazilian, and Bolivian Amazon). "Touch it.... Doesn't it feel like a relative has died?" (*Wars* 32, my ellipsis) said Chota to Berrospi.

In terms of the anthropology of ontologies, this scene provides two interpretations that contribute to my argument about ontological commitments as the main motivation for Indigenous ecological activism. First, by equalizing the logged trees with dead family members, Chota endows the elements of nature with a degree of humanness that subverts the sense of humanness conventionalized by Western rationality and also reiterates the limitations of the notion of cultural diversity by replacing the paradigm of culture with that of ontological diversity. Second, the physical contact that Chota proposes to Berrospi by touching the trunk demonstrates a communicative dynamic that structures the particular reality in which the Asháninkas and the rainforest coexist in a multispecies society in which the Asháninka embodies only one of several forms of personhood. Through this nonlinguistic form of more-than-human communication, deeply sensorial, haptic, spiritual, and indexed in the local cosmology, the ontological identity of the Asháninka people is constructed. This process differs from the social construction of cultural identity in non-Indigenous settings, where logocentric forms of communication are privileged.

Therefore, trees as sentient entities—or actants—have a degree of humanness that allows them to communicate with other human entities, such as Chota. Besides being linked to Asháninka cosmology, the interaction between these different entities fosters Indigenous ecological activism that is mobilized to confront external threats. However, I propose to read this form of communication between ontologies as a way of doing politics otherwise. This alternative politics reverses the values and reconfigures the composition of traditional stakeholders for modern politics. This cosmopolitical Indigenous assembly makes visible their lack of consensus with the modern political lens regarding the ecological conflict because, while the illegal logging mafias perceive trees as inert objects that should be capitalized through their circulation in the global market, to Chota and the Asháninka community, trees are other forms of humans whose preservation is crucial for the survival of the Indigenous world.[6] Chota's Indigenous ecological activism is not about saving trees as if they were global common goods or part of global ecosystemic nature, to recall Ursula Heise (*Sense of Place*). After all, the latter approaches are grounded in the stream of Western environmentalist discourses. However, Chota fights to save a type of personhood whose presence stabilizes Asháninka's world—that is, their ontological setting and its environment-making connection with nonhuman agents. Ultimately, for the Indigenous activists, the conflict is about ontology, the preservation of their world-making modalities, and their ontological survival.

In the face of the external threat incarnated by extractive industries and state or private apparatuses, the modern institutions meant to promote development, the Indigenous ecological activisms embodied by Acuña and Chota reveal a political tension that can be read from a decolonial critique. The institutions of modernity represent how the epistemological domination that the modernization process—which in this specific case occurred with the expansion of extractive industries throughout the twentieth century—imposed in the Andes and the Amazon relates with the hegemonic rationale previously imposed by European colonialism. Decolonial scholars call coloniality to these ongoing oppressive dynamics which centralized Indigenous labor as a means of setting up hegemonic structures both in the colonial and postcolonial periods (Quijano, "Colonialidad"; Mignolo, *The Darker Side*). From this theoretical framework, I propose to address the question of what the conflict is about. Is it possible that this conflict is on the epistemic decolonization of land, identity, and notions of Indigenous nature? Is it possible to examine current ecological conflicts through the lens of categories like dispossession, displacement, and extractivism, which serve as conceptual frameworks shaping the colonial conditions that Indigenous activism seeks to challenge? In the following section, I address these interpellations by analyzing the decolonial scope of Indigenous ecological activists portrayed in *Guerras*.

Decolonial Scopes of Indigenous Activisms

Zárate historicizes Acuña's ecological struggle within a legacy of exploitation with colonial origins. Cajamarca, a region in the northern Peruvian highlands, has a particular milestone in Peru's colonial historiography:

> From then on [sixteenth century], while Spanish soldiers were pillaging the fold from the New World so their aristocracy could squander

it on silk, porcelain and spices from Asia, Cajamarca became another region where wars, epidemics, slave labour in the mines and mestizaje . . . left a deep scar in the emotional memory of the native population. The metaphor of the beggar sitting on a golden bench became, for generations of Peruvians, an allegory of this plundering, which many would say has a contemporary equivalent: unhappy people watching very happy corporations blow up the land to extract gold. (*Wars*, 79–80, my ellipsis)

In Cajamarca took place the first encounter between the last Inca king Atahualpa and the Spaniard conqueror Francisco Pizarro, who kidnapped Atahualpa and, after collecting a massive rescue of gold, betrayed him by killing him in front of his disciples. Since then, this region became an area of extractive exploitation, and Acuña's activism is located in that geohistorical context of colonial continuity in which an Indigenous woman resists in the twenty-first century the imperial ambitions that threaten her very existence. Thus, I propose to read Acuña's and Chota's anti-extractivist struggle within a framework of decolonial perspective that understands the expansion of capitalism in the Americas as defined by Aníbal Quijano ("Coloniality"): as continued Anglo-European domination of commodities and labor forces whose productive articulation gravitates around capital and the global market. In explaining the overlap between colonialism and capitalism as an intersection that is stressed with the emergence of America in global history, Quijano points out that "capitalism as a system of relations of production, that is, as the heterogeneous linking of all forms of control on labor and its products under the dominance of capital, was constituted in history only with the emergence of America" (551). As part of America, Cajamarca is inserted in this colonial dynamic, which depreciates the landscape and its human and nonhuman inhabitants, commodifying them all.

Acuña's anti-extractivist activism is also anticolonial activism inasmuch as she opposes the expansion of extractivism as the ideological and colonial sediment of capitalist modernity, the basis on which epistemological, political, and economic domination over subalternized populations is structured. Internal colonialism, which in *Guerras* is reflected in the complicity of the Peruvian state with the extractive industries, is just one more branch of the broad and complex system of asymmetries that reproduces coloniality in multiple realms, such as race and culture (Quijano, "Colonialidad"), gender and sexuality (Lugones), sources of knowledge (Mignolo, "The Geopolitics"), the being (Maldonado-Torres), and nature (Alimonda; Cajigas-Rotundo). Acuña, Chota, and Cuñachí, the three characters in Zárate's chronicles, experience forms of oppression that interconnect several of these realms of coloniality. However, my interest in focusing on the Indigenous ecological activisms of Acuña and Chota, since Cuñachí is not precisely an activist, is to elucidate the question of what these activists are facing. That is why I illuminate the modern/colonial/capitalist system, in its extractivist form, as the common enemy that both oppose by claiming sovereignty for their ontological commitments.

Notwithstanding, the explicit consequences of extractivism—dispossession, displacement—are not the only issues with which Indigenous ecological activists are struggling. They also struggle against other senses of coloniality that interfere with their sociabilities. As such, large-scale extractivism represents what Héctor Alimonda defines as the coloniality of nature—that is, as the domination of the biosphere (with its human and nonhuman residents) and its territorial configuration (sociocultural dynamics and ecosystems) under a global hegemonic model of "exploitation of the subaltern space" (22). Moreover, Indigenous ecological activists typically have to deal with what Juan Camilo Cajigas-Rotundo (60–63) defines as "biocoloniality": the dynamics of environmentalization of nature promoted mainly by environmentalist discourses that claim the sustainable use of nature, reinserting it as an exploitable good but within an ecocapitalist framework. The environmental conservationist discourse, in some cases, promotes the displacement of human inhabitants with the goal of creating wilderness. However, this approach also threatens the ontological multispecies coexistence that existentially defines Indigenous communities. Although this particular type of domination is not explicit in *Guerras*, it is present in most environmentalist documentaries produced by Anglo-European directors such as those discussed by Marcone. A broader work on the conflicts that traverse Indigenous ecologies cannot omit the epistemological domination implied by some Anglo-European, and also Latin American, environmental activists that, despite their position as external allies, do not necessarily imbricate horizontally with local cosmologies. On that note, something that also entails the domination of Indigenous ecologies is the ontological devaluation of Indigenous people's own condition of humanity.

Chota used to say that he'd had an Asháninka teacher at secondary school who taught him not to be ashamed of his roots. He also swore that one of his grandmothers had been born in an Amazonian community in Iquitos, but her relatives denied it. What made him angriest was realizing that people—politicians, businesspeople, citizens, his own family—believed, in their heart of hearts, that being Indigenous meant being savage, poor and inferior. (*Wars* 29)

A significant part of the chronicle about Chota portrays the tensions between him and his family members, who rejected their Amazonian Indigenous background. In fact, unlike Chota, who returned to live in an Indigenous community in his adulthood, the rest of his family lived in modern cities such as Iquitos and Lima. The strong pride that nurtures Chota's identity inspires his ecological activism and inserts him in the larger agenda of revalorizing the human status of the Asháninka nation's members. If the resistance against the impact of timber logging in the rainforest is a struggle for the recognition of the ontological connection that links the Asháninka people with the nonhuman entities of the Amazon, in a broader sense, this struggle also includes the full acknowledgment of Asháninka's humanness.

Nelson Maldonado-Torres (246–48) points out that the continuum of colonialism-coloniality naturalized in the Americas the nonethics of war that downgraded the ethical standard with which the humanness of the colonized subjects was perceived to justify the violent and genocidal domination over subalternized individuals. He argues that coloniality implies the radicalization of the nonethics of war to such a degree that the colonized and racialized subjects become less than human and, therefore, expendable, killable subjects or, ultimately, those whose survival is only worthwhile as long as they are useful to the purpose of colonization. The attitude that promotes these nonethics is defined by Maldonado-Torres as "misanthropic skepticism" (245), which, by inferiorizing subaltern individuals, diminishes their human status and ranks them on a subontological hierarchy through a set of colonial modalities such as racism, discrimination, sexual violence, exploitation, slavery, and murder. In short, Maldonado-Torres defines this whole scenario as the coloniality of being, a modality of domination in which the dominated individuals continuously experience a situation similar to that of war conditions.

Guerras points out that the conditions of war to which Indigenous activists are exposed transcend human exceptionalism and imply the destruction of the nonhuman agents that are part of the Quechua and Asháninka collectives. In other words, the misanthropic skepticism that facilitates the violence to which these characters are subjected constitutes one more form of dispossession that defines the practical conditions with which extractivism articulates the expansion of capitalist modernity, without leaving behind its colonial foundation.

The murder of Chota and his three Indigenous companions, the destruction of Acuña's house and the attacks against her family, and the physical exploitation of Cuñachí portrayed throughout the chronicles, exemplify the different instances in which the colonial logic of extractivism inferiorizes the humanness of the Indigenous and peasant subjects of the Peruvian Andes and Amazon, together with their nonhuman peers. This is why Chota's demand for full recognition of their humanness, an intrinsic element of his ecological activism, is at the same time a decolonial demand against the coloniality of being that oppresses the Asháninka nation from different fronts and with different representations of kamári. The demons that threaten Asháninka's subsistence are embodied in all the modalities of coloniality (Shining Path attacks, illegal logging mafias, and racism). Ultimately, the dehumanization of Indigenous groups facilitates the normalization of the productivist discourse sedimented by extractivism, which perceives the Andes and Amazon as inert, cultureless territory, waiting to be occupied, exploited, and capitalized. For this reason, I argue that Indigenous ecological activisms are not only promoted by opposition to the formal institutions of modern colonial capitalism (state, corporations), but also confront the institutionalization of the different modalities of oppression that pave the way for these institutions (racial, sexual, and ontological inferiorization).

Ecological conflicts are about ontologies and coloniality. For that reason, decoloniality is manifested in Indigenous activisms in a form of insurrection that mobilizes Acuña and Chota to unleash their struggles in all possible arenas to preserve their particular ecologies and livelihoods (e.g., blockades, lawsuits, media exposure). *Guerras* not only portrays individual cases of reaction to the imminent climate crisis provoked by extractive industries but also the struggle of local groups to safeguard their autonomy against the different forms of intervention carried out by colonial capitalism. When these disputes involve Indigenous socionatural settings, ecological conflicts are also about the coloniality of specific ontologies (e.g., deforestation). This is the reason why Indigenous cosmopolitical assemblies argue for the horizontal engagement of human and nonhuman actors, and also it is the reason why anti-extractivist activisms, mainly Indigenous ones, are undeniably connected to the legacy of anticolonial struggles in Latin America.

Conclusion

As Niall Binns notes, ecological contexts unavoidably tend to permeate Latin American literature. Ecocriticism, applied to the literature of this geographical area, has the mission to explore the literary representations of nature and ecological imaginaries to engage Latin American literary criticism with the contemporary concerns of the world. For this purpose, works such as those by Jennifer French, Jorge Marcone ("The Stone"), and Gisela Heffes ("Introducción," *Políticas*), or compilations such as *Hispanic Ecocriticism* edited by José M. Marrero Henríquez are crucial. However, as Heffes (19) warns, it is necessary to interrogate to what extent ecocriticism is a valid analytical instrument to explore issues concerning ecology and Indigenous cosmologies in a context of contamination. I suspect that Latin American ecocriticism's limitations do not mean per se its uselessness as an analytical instrument if it goes hand in hand with a theoretical effort to address alternative modalities of interrelation between humans and nonhumans in a specific ecological context.

This work constitutes an analytical effort to complement Latin American ecocriticism, since the analysis of ecological conflicts and Indigenous activists gathered in *Guerras* has not focused on the representation of nature but on the motivations and political rationalities that underlie Indigenous ecological activisms. Colonialism and imperialism meant in Latin America the commodification of nature and the human and nonhuman entities that inhabit its different ecosystems. After that period that was historicized as colonial or, better said, as a continuation of it, coloniality entailed the universalization of Anglo-European perspectives on nature adopted by modern nation-state institutions. The complicity of the Peruvian state in the cases analyzed in this work illustrates this neocolonial dynamic. This hegemonic epistemology ideologically structures many of the anthropocentric activities that nowadays we identify as the source of the current ecological and environmental crises. This is the reason why this work considers it necessary to illuminate Indigenous ecological activisms as conflicts that, by implicating Indigenous territories and beingness, transcend the culturalist discourse and settle in the realm of the ontology. At the same time, it is essential to shedding light on the long-lived anticolonial legacy that inspires the Indigenous actors that the colonial discourse dehumanized and constrained as inert objects of nature.

Notes

1. In this chapter, I do not use the terms "environmentalism" and "ecology" as interchangeable concepts. Regarding the former, I follow the connotation elaborated by the decolonial critique of environmentalism (and its locative referent "environment") as a type of discourse that tends to environmentalize nature, converting it into an atmosphere, which entails a tacit acceptance of separation between humans and nature (see Cajigas-Rotundo), since environments are the locations that societies inhabit. On the latter, ecology is regarded in this work following the integrative notion with which this concept is framed in the ontological revisions of ecological activism—that is, a vision that does not presume the separability between humans and nature (see Blaser, "Reflexiones"), since, from a holistic perspective, in this notion societies *are* ecologies.
2. While in the text I will refer to Zárate's work with its Spanish title, the page numbers cited correspond to the English version translated by Granta Books in 2021.
3. The word "rationality" here refers to the conventionalized sense of Western social sciences that define cognitive paradigms and social and political actions through logical methods of deductive and inductive argumentation. This paradigm does not correspond to Indigenous worldviews as argued by Viveiros de Castro and Descola.
4. I am aware that there is another potential framework of interpretation for this conflict that derives from the methodologies of *object-oriented ontology*, where multispecies coexistence implies not a discussion about who exists and who doesn't, but about how all the objects that make up reality exist independently. However, opening this branch of interpretation would imply going beyond the scope and limits of this chapter. For more information on this field, I recommend consulting Timothy Morton's *Ecology Without Nature* (2007) and *Hyperobjects: Philosophy and Ecology After the End of the World* (2013).
5. Bruno Latour (*Politics* 70–77) refers to as *actants* all those human or nonhuman entities that have the capacity to "act" in a collectivity. Latour defines these more-than-human collectivities as assemblies.
6. It is not my intention to assert that an existential consensus is possible among nonhuman agents; at most, a multispecies coexistence. The problem of consensus in this work adopts the sense given by Mario Blaser regarding the modern vision of Western political sectors vis-à-vis Indigenous activism. It is among the latter that the lack of consensus on the ontological condition of nature leads to situations of conflict.

Works Cited

Alimonda, Héctor. "La colonialidad de la naturaleza. Una aproximación a la Ecología Política Latinoamericana." *La naturaleza colonizada. Ecología política y minería en América Latina,* edited by Héctor Alimonda, CICCUS/CLACSO, 2011, pp. 21–59.

Aparicio, Juan Ricardo, and Mario Blaser. "The 'Lettered City' and the Insurrection of Subjugated Knowledges in Latin America." *Anthropological Quarterly,* vol. 81, no. 1, 2008, pp. 59–94.

Binns, Niall. "Acercamientos ecocríticos a la literatura hispanoamericana." *Anales de Literatura Hispanoamericana,* vol. 33, 2004, pp. 9–13.

Blaser, Mario. "Ontological Conflicts and the Stories of Peoples in Spite of Europe: Toward a Conversation on Political Ontology." *Current Anthropology,* vol. 54., no. 5, 2013, pp. 547–68.

Blaser, Mario. "Reflexiones sobre la ontología política de los conflictos medioambientales." *América Crítica,* vol. 3, no. 2, 2019, pp. 63–79.

Cajigas-Rotundo, Juan Camilo. "Anotaciones sobre la biocolonialidad del poder." *Revista Pensamiento Jurídico,* no. 18, 2007, pp. 59–72.

De la Cadena, Marisol. "Indigenous Cosmopolitics in the Andes: Conceptual Reflections Beyond 'Politics.'" *Cultural Anthropology,* vol. 25, no. 2, 2010, pp. 334–70.

Descola, Philippe. *The Ecology of Others.* Prickly Paradigm, 2013.

French, Jennifer. *Nature, Neo-Colonialism, and the Spanish American Regional Writers.* Dartmouth College P, 2005.

Gudynas, Eduardo. "Extracciones, extractivismos y extrahecciones: un marco conceptual sobre la apropiación de los recursos naturales." *Observatorio del Desarrollo,* no. 18, 2013, pp. 1–18.

Heffes, Gisela. "Introducción. Para una ecocrítica latinoamericana: entre la postulación de un ecocentrismo crítico y la crítica a un antropocentrismo hegemónico." *Revista de Crítica Literaria Latinoamericana,* vol. 40, no. 79, 2014, pp. 11–34.

Heffes, Gisela. *Políticas de la destrucción/Poéticas de la preservación. Apuntes para una lectura (eco)crítica del medio ambiente en América Latina.* Beatriz Viterbo, 2013.

Heise, Ursula K. *Sense of Place, Sense of Planet: The Environmental Imagination of the Global.* Oxford UP, 2008.

Kohn, Eduardo. "Anthropology of Ontologies." *Annual Review of Anthropology,* vol. 44, 2015, pp. 311–27.

Latour, Bruno. *Politics of Nature.* Harvard UP, 2004.

Lugones, María. "Heterosexualism and the Colonial/Modern Gender System." *Hypatia,* vol. 22, no. 1, 2007, pp. 186–209.

Maldonado-Torres, Nelson. "On the Coloniality of Being: Contributions to the Development of a Concept." *Cultural Studies,* vol. 21, no. 2–3, 2007, pp. 240–70.

Marcone, Jorge. "Filming the Emergence of Popular Environmentalism in Latin America." *Global Ecologies and the Environmental Humanities: Postcolonial Approaches,* edited by Elizabeth DeLoughrey, Jill Didur, and Anthony Carrigan, Routledge, 2015, pp. 207–25.

Marcone, Jorge. "The Stone Guests: Buen Vivir and Popular Environmentalisms in the Andes and Amazonia." *The Routledge Companion to the Environmental Humanities,* edited by Ursula K. Heise, Jon Christensen, and Michelle Niemann, Routledge, 2017, pp. 227–35.

Marrero Henríquez, José Manuel. *Hispanic Ecocriticism.* Peter Lang, 2019.

Mignolo, Walter. *The Darker Side of the Renaissance. Literacy, Territoriality, and Colonization.* U of Michigan P, 1995.

Mignolo, Walter. "The Geopolitics of Knowledge and the Colonial Difference." *South Atlantic Quarterly,* vol. 101, no. 1, 2002, pp. 57–96.

Quijano, Aníbal. "Colonialidad y Modernidad/Racionalidad." *Perú Indígena,* vol. 13, no. 29, 1992, pp. 11–20.

Quijano, Aníbal. "Coloniality of Power, Eurocentrism, and Latin America." *Nepantla: Views from South,* vol. 1, no. 3, 2000, pp. 533–80.

Santos-Granero, Fernando. "The Enemy Within: Child Sorcery, Revolution, and the Evils of Modernization in Eastern Peru." *In Darkness and Secrecy: The Anthropology of Assault Sorcery and Witchcraft in Amazonia,* edited by Neil L. Whitehead and Robin Wright, Duke UP, 2004, pp. 272–305.

Todd, Zoe. "Indigenizing the Anthropocene." *Art in the Anthropocene: Encounters Among Aesthetics, Politics, Environments and Epistemologies,* edited by Heather Davis and Etienne Turpin, Open UP, 2015, pp. 241–54.

Viveiros de Castro, Eduardo. "Cosmological Deixis and Amerindian Perspectivism." *Journal of the Royal Anthropological Institute,* vol. 4, no. 3, 1998, pp. 469–88.

Viveiros de Castro, Eduardo. "Exchanging Perspectives: The Transformation of Objects into Subjects in Amerindian Ontologies." *Common Knowledge,* vol. 10, no. 3, 2004, pp. 464–84.

Zárate, Joseph. *Guerras del Interior.* Debate, 2018.

Zárate, Joseph. *Wars of the Interior.* Translated by Annie McDermott. Granta, 2021.

5

Archives of the Planetary Mine

Art, Political Ecology, and Media Geology in Chile and Venezuela

GIANFRANCO SELGAS

Introduction: Mines, Minerals, Media

The relationships humans establish with the Earth and its strata are integral to the social relationships of labor and exploitation characteristic of nineteenth-century industrial capitalism and the contemporary financial and digital capitalism of the twenty-first century.[1] From mineral extraction and energy consumption to the primarization of nature and the production factories of computing equipment, disentangling the political from the natural and the geopolitical from the geological is increasingly complex. In current discussions and debates on the Capitalocene (Moore 169–70), the geological plays a determining role for capitalism and the abstraction of nature and labor into value.[2] This extends beyond how Western geological sciences and stratigraphic records have shaped our understanding of the earth and its functioning. It also encompasses how the Earth's components, in direct relation to racial and colonial legacies establishing differences between organic and inorganic life, initiate a redefinition of what it means to be human, thereby influencing the formation of ideologies, history, and politics.

As Adam Bobbette and Amy Donovan (2–4) observe, the appropriation, ordering, and distribution of fossil fuels, minerals, and land underscore the intrinsic relationship between geological and political formations, giving rise to the concept of political geology. This concept highlights the interconnectedness of the geos and the bios, the inorganic and the organic, the lifeless and the living, as these relationships are experienced and debated on material, cultural, political, and geographical levels. Kathryn Yusoff emphasizes that identifying humans and capital as geological forces that alter the planet's ecological balance underscores the intricate material and symbolic exchanges and entanglements between geological strata and social worlds.

In the Latin American context, these exchanges are particularly prominent. The exploitation of the subsoil and mineral deposits has significantly influenced the region's economy for over five centuries. As Kendall Brown (128–29) explains, this exploitation involved both structural and cultural changes. The modernization of the mining industry, driven by foreign capital in the early twentieth century, introduced new technologies for geological exploitation and led to the proletarianization of labor. Foreign and national capital clashed over territorial and productive control of the mines, while companies faced growing resistance from union organizers over workplace control. The subsequent proletarianization of mining labor transformed the economic and cultural landscape of mining areas. This transformation combined the use of heavy machinery, explosives, and open-pit mining techniques with forced and waged labor to exploit various types of ores. The adoption of new technologies, machinery, and refining plants, along with innovations in chemical and synthetic processes, encountered cultural resistance. Companies attempted to overcome this resistance by importing foreign managers and technicians, whose higher salaries and better housing provoked reactions from the locals.

Mines and minerals can be understood as spaces and materials that unify local and planetary histories. Martín Arboleda (4–13) explains that mineral extraction involves a global network of production, material exchange, and technologies that connect nations and materialities across time and space. A mine is not merely a discrete sociotechnical object but a dense network of territorial infrastructures and spatial technologies that expand globally due to the multiple uses of minerals in supporting modern lifestyles. Emphasizing this aspect highlights the relevance of mineral exporting nations such as Chile and Venezuela. These countries are configured in the global financial circuit as extractive nations of the Global South and are essential for understanding the multiple mechanisms of the commodification of nature and supply chains.

Building on this perspective, I explore how contemporary Latin American art symbolically and materially addresses the exchanges and entanglements between geology, politics, and culture. I develop the notion of "archives of the planetary mine" to conceptualize art as a cultural archive that records the socioecological entanglements activated by the multifaceted dimensions of mining in Latin America. Through two scarcely studied works by Venezuelan artist Ana Alenso (Caracas, 1982), I examine the relationship between geology, politics, and culture. Alenso's work, which combines sculpture, photography, installation, sound, and video, is informed by intense research supported by scientific studies and reports, as well as exchanges with environmental activists. Her art offers allegorical reflections on the economic, social, and ecological risks and conflicts associated with nature extraction mechanisms in Latin America and the configuration of petrocultural imaginaries in Venezuela.

Although Alenso's artistic work has not yet received detailed attention from specialized critics, it can be situated within the context of contemporary Latin American artists engaging with what Macarena Gómez-Barris (1–9) describes as the "extractive zone." This term refers to regions of extractive capitalism where the violence inherited from colonial exploitation reduces, constrains, and commodifies life. The art of the extractive zone explores the dilemmas of pressing socioecological crises at both local and planetary levels from decolonial, queer, and submerged perspectives and epistemologies. Other relevant artists in this context include Cecilia Vicuña, Esperanza Mayobre, Carolina Caycedo, Nohemí Pérez, and Andrea Canepa, to name but a few. In this chapter, I aim to situate Alenso's work in this context, focusing specifically on her film *Desviar la inercia* (Diverting inertia, 2019), centered on the copper and lithium extractive zone in Antofagasta, Chile, and her multimedia installation *Lo que la mina te da, la mina te quita* (What the mine gives, the mine takes, 2020), which examines the extraction of gold and coltan in the Amazonian and Guayana region of Bolívar state, Venezuela.

In what follows, I discuss how *Desviar la inercia* and *Lo que la mina te da, la mina te quita* articulate claims and actions that discuss extractive capitalism in Chile and Venezuela. To this end, I follow Gabriela Merlinsky and Paula Serafini's assertion that art, as a creative and communicative vehicle of collective narratives and identities, enables different experiences and social activism to propose "nuevos modos de vida para oponerse a formas de naturalización que niegan la crisis ambiental y que asimismo promueven formas de silenciamiento en torno a las consecuencias del extractivismo [new modes of living to oppose forms of naturalization that deny the environmental crisis and also promote forms of silencing around the consequences of extractivism]" (16). According to T. J. Demos (8), environmentally engaged art has the potential to rethink politics and politicize art's relation to ecology, highlighting how nature is inextricably linked to economics, technology, culture, and law at every turn.

As I will indicate later, Alenso's artistic practice can be analyzed as an artistic praxis committed to the socioecological crisis, arguing that it is part of an aesthetic and political redistribution of the sensible, providing a different sensorium to create new material rearrangements of signs and images (Rancière 39). Her work questions, in an experimental, imaginative, and radically thoughtful sense, the patterns of colonization, appropriation, and exploitation of the web of life (Merlinsky and Serafini 15–20). By highlighting the complex relationships woven in specific local spaces—the mining areas of Antofagasta and Bolívar—with the global formations that drive them, her artistic pieces offer "new ways of comprehending ourselves and our relation to the world differently than the destructive traditions of colonizing nature" (Demos 19).

In what follows, I draw on the film and installation to explore how they are intrinsically related to the planet's history, geological formations, and minerals. Alenso's work deals symbolically and materially with the various historical-geographical, geological, and political magnitudes of energy consumption and mining as a condition of possibility for both technological updating and cultural production itself. To address this, I draw on Jussi Parikka's (21–23) work on the social and environmental importance of various minerals in understanding the material, technological, and cultural development of modern life. Discussing media geology is key to understanding the geological in relation to, but also beyond, the interior and exterior strata and folds of the earth. It invites us to think of the geological as a way of interrogating its living materiality and decentering categories that isolate humans on a biodiverse planet.

Metals such as aluminum, copper, gold, and iron, as well as minerals like lithium, iridium, cerium, tantalum, tungsten, thorium, and manganese, have been essential elements for understanding industrial, technological, and digital developments from the nineteenth to the twenty-first centuries. They also highlight the intrinsic relationship between media materiality and the geology of mining and metals that enable the constitution and functioning of the technological apparatus that envelops modern life. This approach delineates and clarifies how the organic and inorganic integrate into tangible assemblages, materialized in daily practices involving, for example, the use of copper in transportation vehicles and cell phones or the speculative value of gold as a financial hedge against inflation and economic interest rates. According to Parikka (26, 141), the subsoil and the earth must be understood as another type of medium that communicates and is communicated. In other words, talking about media geology implies tracing the relationships between geological strata and media materialities, where the histories of extractive capital, labor, and the planet are entangled in the technological and cultural devices that constitute them.

Finally, I conclude by conceptualizing Alenso's selected works as pieces of what I call "archives of the planetary mine." I outline this concept to describe ecologically and politically oriented artistic praxis as a form of cultural archive that records the socioecological dialectic activated by the multiple dimensions of mining's impact in Latin America.

Desviar la inercia: Copper, Visibility, and Deterritorialization of the Subsoil

With the decline of the importance of Chilean nitrates in the 1930s, the development of large copper mines by US capital reoriented the country's economy from Europe toward the United States and, later, Asia. The central and northern regions of Chile, particularly the territories of Antofagasta and Chuquicamata surrounding the Atacama Desert, possess large deposits of copper and lithium. During the twentieth century, the Chuquicamata open-pit mine produced more copper than any other mine in the world, and, currently, 96 percent of the exports leaving the ports of Antofagasta are various types of minerals (Brown 132; Arboleda 75–76). In *Desviar la inercia,* Alenso visits these mining regions of Chile to explore the material and social infrastructures of mining extractivism in the country. The film, developed in the context of the SACO 8 Contemporary Art Festival held in Antofagasta, is divided into two parts. The first part, with an aesthetic, experimental, and documentary orientation, records various shots of the desert environment and the mining infrastructures of Antofagasta (e.g., mines, heavy machinery, smelters, trains, ports) interspersed with texts and informative comments on mining in Chile. The second part, properly documentary and testimonial, collects interviews with scientists, academics, activists, residents, and workers of Antofagasta who share collective reflections and warnings around the extraction of natural resources.

Desviar la inercia makes visible the intrinsic relationship between the materialities of the subsoil and modern daily life, and the symbolic and material components that interlace them in a local and planetary context shaped by the mining and its supply chain. The first shot of the film is representative of this. It begins with a close-up of a Formica table with blue and green hues, while a thunderous background sound, metallic and electrical,

Figure 5.1. Industrial facility in Chile. Screenshot. *Desviar la inercia,* by Ana Alenso.

serves as a sound effect. The camera moves over the table's surface, giving the impression of an aerial view over the sea while the following text appears on screen: "A la orilla del mar flotan finísimas partículas de metales pesados, imperceptibles, letales [At the edge of the sea, very fine particles of heavy metals float, imperceptible, lethal]" (Alenso, *Desviar* 00:10–00:26). The camera's movement over the table reveals the mundanity of the space: a coffee maker, an electric kettle, and a sink, as well as a clock hanging on one of the room's walls. When the camera widens the shot, it shows the space of a small kitchen or coffee area, possibly in a studio or urban office.

The roar accompanying the film's first seconds, however, does not seem connected to the everyday kitchen space. It is only a few seconds later, with a change of shot, that the full frame of an industrial facility in operation is shown, with the sea in the background, presumably where the industrial processing sounds that start the video are emanating from (see fig. 5.1). The correspondence established between both images, initially disconnected by representing two nonoverlapping and opposite spaces, is marked by the imperceptible relationships that, as indicated by the text in the previous scene, link the everyday with the contamination of land and water due to the extraction processes associated with the mining industry. Here, sound plays a predominant role, establishing an almost imperceptible visual connection between the home space and the mining extraction process. Although that world of metals, metalloids, and minerals cannot be seen at first glance, the

Figure 5.2. Degraded geography and Chuquicamata. Screenshot. *Desviar la inercia,* by Ana Alenso.

visual-auditory interplay set by the video allows connecting, through the industrial roar that sneaks into the kitchen space, the Formica table, and everyday appliances, the invisible relationship established with the mining infrastructure.

This relationship between the invisible and the auditory recurs throughout the first part of the film, gradually making explicit the imperceptible connections where the subsoil, minerals, and humans are more interconnected than they initially seem. One of the most striking examples occurs shortly after the previously discussed scene, when a cut reveals a kilometer-long stretch of mountain or rocky hill, seemingly degraded by mining activity. The desert environment of the foothills, resembling a sea of endless land, challenges the notion of lifelessness—that is, viewing the rocky, arid, and inorganic as something dead or inert—by superimposing the names of various mining business groups, both national and transnational, as well as industrial and technological service companies, and different copper deposits located in the Antofagasta area. This allows for rethinking the supposed inert and lifeless condition of the geological territory as a space where the existence of metals and minerals elicits a series of geopolitical and economic interests that are rendered invisible in the apparent monotony of the Chilean desert. Thus, the image of the rocky and repetitive environment calls for reinterpretation based on its added value for national and foreign capital when the names of mining companies begin to appear incessantly, all in golden letters. The list

Figure 5.3. Gas and copper. Screenshot. *Desviar la inercia,* by Ana Alenso.

of companies with mining exploitation purposes is interspersed with the mention, also in golden letters, of copper-rich deposits, as well as the various metals and minerals extracted in the area (see figs. 5.2 and 5.3).

The systematic and almost mechanical mention of the names of companies, deposits, minerals, and metals superimposed on the product of the extractive enterprise—as observed in figures 5.2 and 5.3, the degraded geography or the gas expelled into the atmosphere—illuminates the relationship, at first glance imperceptible, of the ecology of capitalism. In other words, the valuable geological formations of Antofagasta mobilize transnational and national companies to transform the territory through the processes of extraction, processing, and export of ores converted into raw materials. At the same time, this also maps the different subsoil elements that integrate the commodities supply chain and are fundamental to the development of modern life. This, in Parikka's words, implies a way of rethinking the subsoil as a medium that "connects to the wider geophysical life worlds that support organic life as much as technological worlds" (4).

On the other hand, the film highlights a "deterritorialization" (Deleuze and Guattari 441) of Antofagasta, where different bodies and techniques are pressured to act and behave according to the spatial configuration of extractivism. To deterritorialize Antofagasta implies abstracting it from its natural condition and turning it into a value category. In the *Grundrisse,* Marx (510, 634) sketches out how the abstraction-extraction logic works through the combination of a set of productive forces under the impulse of capital to integrate material substrata into society. Driven by the production of value, capitalism abstracts the plurality inherent in the mines and the planet by deterritorializing them under forms of enclosure, extractive zones, and profit and wealth accumulation. Capital does not distinguish between mines or recognize labor beyond its role in extraction and profitability. Instead, it reduces labor to an extractable singularity, transforming both organic and inorganic life into surplus value and commodities.

This is evident in the examples discussed above, and it becomes more apparent in a scene featuring a citation from the micropolitics of Guattari and Suely Rolnik, reflecting on the historical and geographical becoming of this northern Chilean region: "El territorio se puede desterritorializar, esto es, abrirse en líneas de fuga, y salirse de su curso, hasta destruirse [The territory can be deterritorialized, that is, opened in lines of flight, and go off course, to the point of self-destruction]" (Alenso, *Desviar* 02:12–02:25). The quote reflects on the environmental impact of mining and the emergence of infrastructures as a catalog of deterritorialization activated by capitalism's abstraction-extraction logic and the sociotechnical infrastructures that constantly pump minerals toward the ports of Antofagasta, traversing an arid and fractured urban landscape where wealth and poverty coexist. This form of deterritorialization of the subsoil implies "the functional integration of the supply chain of extraction, rendering its own spatial register through an expanding yet fragmented and geographically uneven fabric of urbanization" (Arboleda 106).

This pattern responds to a historical trend in Chilean mining zones. As Brown (133) explains, geography has always been a problem for mining exploitation in the country, especially in the northern mines, such as Chuquicamata. The mining and milling techniques developed throughout the twentieth century required large machines to handle massive amounts of ore and were more capital intensive than labor intensive. US companies leading the process needed fewer semi-skilled permanent workers rather than the large number of unskilled transient workers required in earlier times. One solution was for the company to build towns to house its workers, providing them with a place to live while creating dependence on the company for subsistence and housing. These aspects are salient in the film, especially through the continuous recording of areas converted into industrial spaces, which have come to be called sacrifice zones: geographic spaces with high industrial concentration where priority is given to establishing industrial hubs at the expense of social and environmental well-being.

This notion, however, is explicitly criticized by the film, which dismantles the conceptualization that subjugates these regions to the abstract constructs of mineral and human exploitation for capitalist benefit or the production of value. As stated in the film: "¿Zonas de sacrificio? Yo prefiero llamarlas zonas de resistencia. Una resistencia bajo la luz del sol nortino, una resistencia científica, ética. Una resistencia de testimonios y arraigos [Sacrifice zones? I prefer to call them resistance zones. Resistance under the northern sun, scientific resistance, ethical resistance. Resistance of testimonies and roots]" (Alenso, *Desviar* 06:24–06:36). This marks the transition to the second part of the film. While the first half illuminates the material correlation between the organic and inorganic, the second part offers another form of visibility, this time through the successive testimonies of members of the scientific community and local residents who live in the region. *Desviar la inercia* highlights the historical characteristics of mining under capitalism as a sociotechnical arrangement of the planet. As the film's title suggests, artistic praxis can serve to divert the inertia of the deterritorialization of the subsoil, making visible the intricate relationship between copper, minerals, and labor/humans in the web of life.

Lo que la mina te da, la mina te quita: Gold, Deep Temporality, and the Materiality of Extraction

Covering an area of 112,000 square kilometers, the Orinoco Mining Arc (OMA) is a strategic mining zone designated to exploit deposits of gold, diamonds, coltan, and other minerals in the Venezuelan Amazon and Guayana region. It currently represents 12 percent of the country's territory. Established by executive decree on February 24, 2016, the OMA exemplifies the socioenvironmental impact produced by development policies and attempts to find solutions to a declining oil economy. According to Antulio Rosales (1312–13), the crisis of oil prices in 2014 and the imposition of international sanctions on Nicolás Maduro's administration led the Venezuelan state to implement a "radical rentierism." Rosales describes how the state has progressively expanded its extractive frontiers from oil and gas to intensive mining and the extraction and issuance of cryptocurrencies backed by crude oil and gold, aiming to exploit mineral elements of the subsoil connected to global financial circuits. This radical rentierism has transformed the traditional rentier state, shifting from a structure seeking to maximize economic benefits through royalties and taxes on oil companies to one that expands and deepens the sources of rent capture in a process of appropriation and exploitation of the subsoil before and during the process of material extraction.

Historically, the Venezuelan Amazon and Guayana region have alternated between neglect and attention from the national and regional government. Although the OMA has its precedent in the opening of the Imataca Forest Reserve, in the jurisdiction of Bolívar and Delta Amacuro states, decreed by Rafael Caldera's government in 1997, the exploitation project was resumed during Hugo Chávez's administration and executed by Maduro. Although official statements refer to the OMA as "un modelo de minería responsable [a model of responsible mining]" (Ministerio del Poder Popular de Desarrollo Minero Ecológico), the environmental, social, and ecological costs of this mega-mining enterprise are dire, especially in terms of ecosystem degradation, health crises from the use of toxic substances, the flourishing of illicit businesses involving mafias, military, and guerrillas, as well as systematic violations of human and environmental rights, and sovereignty issues in southern Venezuela (SOS Orinoco "Illegal Mining"; Terán Mantovani).

In the multimedia installation *Lo que la mina te da, la mina te quita*, Alenso explores the mechanisms and consequences of gold, coltan, and diamond mining in southern Venezuela. In 2020, the installation was exhibited in Berlin, Germany, consisting of two main pieces made from material and industrial waste linked to mineral extraction: a *tame* (sluice box) and a mining work cabin—I will detail and analyze these pieces shortly. In various interviews, the artist has spoken of this installation as a speculative one (Mustroph) where "signos de autodestrucción y como un modus operandi antropocéntrico [signs of self-destruction and an anthropocentric modus operandi]" are revealed, aiming to denounce an ecocide and "las contradicciones e incertidumbres provocadas por las turbulencias económicas y la permanente exportación de recursos se trasladan al espacio expositivo estética y hápticamente en una máquina escultórica [the contradictions and uncertainties caused by economic turmoil and the permanent exportation of resources are transferred aesthetically and haptically to the exhibition space in a sculptural machine]" (Alenso, "Lo que la mina te da"). By revealing unconventional evidence in the media about the alliances between mining, violence, money, and power, the installation shows the complicities between the economy, extractivism, and death.

However, a scarcely discussed aspect of this installation is how it draws attention to the very materiality of extractivism—that is, how the piece allows a reflection on the chemical and material processes that come into contact during the gold extraction process. This is notable in the sluice box piece (see

Figure 5.4. Artisanal sluice in the installation *Lo que la mina te da, la mina te quita*, by Ana Alenso. Photo by Ana Alenso, 2020, *Lo que la mina te da, la mina te quita*.

fig. 5.4). The sluice box is an open, inclined channel where the fluid material passes so that the particles of gold and mercury are trapped (Lozada 468). In the extractive zones of the Venezuelan Amazon, miners use these tools to obtain gold directly. The procedure, however, is highly contaminating and nonbiodegradable. Pressurized water is used to break the soil and extract gold-bearing alluvium. Using a motor-pump system that discharges pressurized water jets, they break the soil while another suction system collects the loosened material and transports it to the sluice box. For the process to be effective and obtain gold, miners need to add mercury to the suction barrels. The water then flows over a ramp covered with an aluminum plate, which magnetically attracts the mercury. When the water pumping stops, the amalgamated particles of gold and mercury adhere to the aluminum plate, obtaining the mineral benefit while returning the contaminated water to the river.

In Alenso's piece, the chemical process of mining is exposed. Just as the chemical interaction between mercury, soil, water, gold, and aluminum results in the sieving of the coveted metal and the contamination of the water, the installation also experiments with the infrastructure of mining extractivism by transposing the assemblage of the extractive apparatus into the exhibition space. Beyond discussions about the processes and forms of autonomization of art in exhibition and museum spaces, I am instead interested in thinking of this artistic praxis as another form of deterritorialization, this time rehearsing a reverse movement to the deterritorialization of the subsoil activated by capitalism, as discussed in the previous section. Following Parikka, "the sciences and the arts often share this attitude of experimentation and the experiment—to make the geos expressive and transformative" (57). The operation initiated by the piece redirects the focus toward aspects often neglected in discourses of geopolitics and economic policy; in other words, how the geos, the earth, the soil, and the Earth's crust, the rivers, chemical processes and machines, technology, and human labor act in unequal assemblages to convert the earth into an extended resource.

This also highlights another way of thinking about time and space, as the installation emphasizes recognizing the radical temporality of the Earth, which does not conform to the linear and narrative time corresponding to the historical development of modern societies. By emphasizing the chemical process of gold extraction and how various material assemblages intersect in mining labor, the installation's sluice box shifts the hermeneutic focus that characterizes interpretations of the human world to one of material order—in the embodied sense of things. In other words, the assembled materials resonate with a political geology that informs us of other dynamics of relationship (human and geological) and temporalities (the millennial cycles in the formation of minerals and the rapidity of their commodification in the financial circuit) that condense in extractive activity.

Alenso's installation allows us to consider the deterritorialization of the materiality of extractivism to explore the various relational processes that emerge in mining spaces. Additionally, the sluice box establishes a fundamental connection between the relationships of the deep temporality of the Earth, which began to gain interest in geological study circles during the eighteenth and nineteenth centuries, and the new mappings of geology and natural resources that geopolitics and political economy trace based on financial, military, and energy interests (Parikka 43–44). I understand this deep temporality as linked to geological timescales of decay and natural renewal, but also to the multiple historical and current dynamics of capitalist interests and socioecological crisis. The sluice box operates as a tool of material and symbolic order: on the one hand, it transports the infrastructure of mining extractivism from the Amazonian territory to the European gallery, making visible how it operates; on the other hand, it invites reflection on the relationships established between mineral extraction and modes of human life, highly dependent on these processes of environmental predation. Put differently, the sluice box allows us to observe and think about the ways in which intertwined geology, politics, and culture are relevant to rethink how we are involved in, and part of, the planet's transformation cycles as mediated by plundered dynamics.

Moreover, *Lo que la mina te da, la mina te quita* also urges us to consider how the artistic medium illuminates the inherent contradictions in the current discourse focused on energy transition. This becomes clearer in another key piece of the installation: the mining work cabin (see fig. 5.5). The cabin features three screens inside showing satellite images of the areas destroyed by mineral extraction, individuals working intensely in illegal mines flourishing in the southern Venezuelan region, and testimonies of various people involved, interested, or directly and indirectly affected by what happens in the OMA. Minerals such as gold, copper, lithium, or coltan are essential components for the proper functioning of various current digital devices, but, in recent years, their market value has also significantly increased due to their important role in crafting systems for energy transition, as an attempt by nations of the Global North to reduce carbon emissions from the consumption of fossil and nonrenewable energies. While a radical change in energy consumption policies and in the capitalist structure governing them is imperative, the energy transition risks being anything but fair without offering cultural and structural changes in supply chains and in the governance of extractive industries (Riofrancos; Szeman and Diamanti).

The mining cabin can be interpreted as a reference to the incoherence of the discourse of green capitalism and energy transition made possible through new forms of energy acquisition focused on mineral extraction such as coltan, lithium, and copper. By situating within the mining infrastructure the contemporary history of destruction that it itself sustains, the piece permeates the historical background of extractive practice in Venezuela, anticipating its future socioenvironmental impact. Seen in this way, the installation dismantles the illusory idea that has conditioned the country, first as a quintessential oil nation and second as a developing mining country. This conception, described by Fernando Coronil (387–92) as the naturalization of wealth and the idea that wealth exists directly in nature, is dismantled in the piece by questioning the illusion of oil and minerals as materials with magical properties that, once inserted into the international market, transmute into modernization, exchange value or money, technology, and (re)insertion into the global financial circuit.

Figure 5.5. Mining cabin in the installation *Lo que la mina te da, la mina te quita*, by Ana Alenso, at Galerie Wedding, Berlin, 2020. Photo by Tomás Eyzaguirre.

Taking the above into consideration, the piece visualizes the dismantling practice of this illusion to begin thinking about Venezuelan and planetary history in a dynamic that concerns socioecological relations with geological and mineral formations and our daily interaction with energy. This last aspect, extensively reflected upon by Alenso ("Lo que la mina te da"), is expressed through the phrase "We are gold and diamond dust" projected on one of the screens of the mining cabin (see fig. 5.5). Alenso ("Editor's Note" 8) links the phrase to Amazonian Indigenous cosmovision and their alternative ontologies to experience the world. As they offer a different coexistence with nature and the gold mines, the phrase deploys another perspective to that conveyed by capitalist abstraction and extraction, rethinking the territory not only from an ecological or anthropological perspective, but also from an existential point of view. The phrase is an invitation to take seriously the inherent relationship forged between the human and capitalist civilizational model and the minerals that enable and sustain both the functioning of modern energy systems and human consumption patterns. With this, the message is relevant because it explicitly exposes the mineral, human, and technological entanglement in the web of life set in motion through the extractive work cabin, illuminating the geomaterial source of the matter to unveil the essence that is part of the "magic act" of the illusory.

I understand "the magical" according to the conceptualization proposed by Coronil (387–88). His starting point is the Hegelian premise that the West has defined its position toward nature as that of a "maker of miracles." Humanity, understood as a "miraculous worker," subordinates nature as the impotent material of its own activity. However, in the "production of miracles" of Venezuela's radical rentierism, what really happens is a symbolic change from oil and gold to money, transfiguring the social agents involved in that process. Following Coronil's thesis, while the state has presented itself as a miracle maker that could turn its domination of nature into a source of historical progress, the reality is that the state has been limited to producing increasingly precarious and scarce "acts of magic," since its power only comes from the powers of mineral money subject to fluctuations in the international market. By paying attention to this, *Lo que la mina te da, la mina te quita* confronts us with the social and environmental consequences connected to nature as a totality in Venezuela, as a flow of flows where human beings make the environment, and the environment makes human beings (Moore 3–4). Alenso's work not only engages with the modern energy imaginary involving the "acts of magic" that "put out of sight" fossil fuels and minerals, transmuting them into material progress, value, and modernization. Instead, it allows us to see beyond the abstraction of the geos into value and understand it from an alternative perspective considering the entanglement of mineral matter and geological forces with social practices.

Conclusion: Archives of the Planetary Mine

Desviar la inercia and *Lo que la mina te da, la mina te quita* by Venezuelan artist Ana Alenso illuminate how art enhances socioecological claims and actions that discuss extractive capitalism in Chile and Venezuela. I highlighted how these pieces underscore the political and economic history of copper and gold exploitation as more than mere exchange value commodities. The film highlights the historical characteristics of mining under capitalism as a sociotechnical arrangement of Antofagasta. As the film's title suggests, artistic praxis can serve to divert the inertia of the deterritorialization of the Chilean subsoil, making visible the intricate relationship between copper and labor/humans in the web of life. On the other hand, the installation exposes the mineral, human, and technological entanglements set in motion through the extractive work cabin and the sluice box, illuminating the geomaterial sources of capitalism and everyday life to unveil and see beyond the abstraction and extraction dynamic of value production.

Art maintains a close relationship with the planet's natural history. Understanding the material relationships conveyed by video art, in a symbolic sense, and multimedia installation, primarily in a material sense, is crucial for grasping how artistic practices can illuminate the political ecology of extractivism. In the film, humans and the geos, often overshadowed by profit and commodity exchange value, become visible through the naming of minerals and the voices of communities expressing their discontent with the mining industry. In the installation, this visibility is achieved through the materials used to construct the extraction apparatus, connecting the spectator with the deep temporality of the earth and the apparatus of extraction.

Based on these points of reflection, I consider that these artistic works can be conceptualized as "archives of the planetary mine." If, according to Arboleda, the notion of the planetary mine implies "one that vastly transcends the territoriality of extraction and wholly blends into the circulatory system of capital, which now traverses the entire geography of the earth" (5), then its archive must be one that manages to register, in various formats, the dilemmas and problems imposed by such connectivity. These artistic expressions can be thought of as a form of cultural archive that record the socioecological and spatial imbrication activated by the multiple dimensions of mining and other extractive practices. As I have argued elsewhere (Selgas 1018–19), the symbolic and material potential of minerals, as the region's history is completely entangled with these materialities and capitalism, serves Alenso as a tool to explore the past, present, and future. This invites us to think more carefully about the articulation between symbolic imagination and the materiality of extractive processes in Latin America, the Global South, and the rest of the world, addressing the epistemological, historical, political, cultural, and geological displacements and shifts activated by art as a form of political ecology.

Notes

1 This chapter is an English translation, with minor modifications, of its original publication in Spanish: Gianfranco Selgas, "Archivos de la mina planetaria: Arte, ecología política y geología de medios en Chile y Venezuela," *Diálogos Latinoamericanos,* vol. 31, 2022, pp. 110–25, doi:10.7146/dl.v31i.132589. Licensed under Creative Commons CC-BY 4.0.

2 The Capitalocene refers to the forced coercion of human and nonhuman labor subjected to the unlimited accumulation of capital, which has disrupted the balance of the planetary ecosystem and the human-nature metabolism. This crisis is caused by a sector of the world population that is closely tied to the exploitation of nature and the deepening of capitalism as a world-system. For a conceptualization of the Capitalocene as an alternative to the Anthropocene, see Jason W. Moore (169–93).

Works Cited

Alenso, Ana, director. *Desviar la inercia*. Instituto Goethe Chile and SACO 8. https://vimeo.com/350974999.

Alenso, Ana. "Editor's Note." *Lo que la mina te da, la mina te quita / What the Mine Gives, the Mine Takes*, edited by Ana Alenso, Bom Dia, 2023, pp. 5–8.

Alenso, Ana. "*Lo que la mina te da, la mina te quita*." *Artishock: Revista de arte contemporáneo*, 2021, https://artishockrevista.com/2021/01/14/ana-alenso-lo-que-la-mina-te-da-la-mina-te-quita/.

Arboleda, Martín. *Planetary Mine: Territories of Extraction Under Late Capitalism*. Verso, 2020.

Bobbette, Adam, and Amy Donovan. "Political Geology: An Introduction." *Political Geology: Active Stratigraphies and the Making of Life*, edited by Adam Bobbette and Amy Donovan, Palgrave Macmillan, 2019, pp. 1–34.

Brown, Kendall W. *A History of Mining in Latin America. From the Colonial Era to the Present*. U of New Mexico P, 2012.

Coronil, Fernando. *The Magical State: Nature, Money, and Modernity in Venezuela*. U of Chicago P, 1997.

Deleuze, Gilles, and Félix Guattari. *A Thousand Plateaus. Capitalism and Schizophrenia*. U of Minnesota P.

Demos, T. J. *Decolonizing Nature: Contemporary Art and the Politics of Ecology*. Sternberg, 2016.

Gómez-Barris, Macarena. *The Extractive Zone: Social Ecologies and Decolonial Perspectives*. Duke UP, 2017.

Lozada, José Rafael. "Opciones para una minería de oro que cumpla con las normas ambientales, en la Guayana venezolana." *Revista Geográfica Venezolana*, vol. 58, no. 2, 2017, pp. 464–83, https://www.redalyc.org/articulo.oa?id=347753793012.

Marx, Karl. *Grundrisse: Foundations of the Critique of Political Economy*. 1857–58. Penguin, 1993.

Merlinsky, Gabriela and Paula Serafini. "Introducción." *Arte y Ecología Política*, edited by Gabriela Merlinsky and Paula Serafini, Universidad de Buenos Aires and CLACSO, 2020, pp. 11–26.

Ministerio del Poder Popular de Desarrollo Minero Ecológico. "Arco Minero del Orinoco (AMO): un modelo de minería responsable." *Ministerio del Poder Popular de Desarrollo Minero Ecológico*, n.d., https://www.desarrollominero.gob.ve/?page_id=59652.

Moore, Jason W. *Capitalism in the Web of Life. Ecology and the Accumulation of Capital.* Verso, 2015.

Mustroph, Tom. "Gold, Nickel, Diamanten und jede Menge Gift." *Neues Deutschland,* 2021, https://www.nd-aktuell.de/artikel/1146867.gold-nickel-diamanten-und-jede-menge-gift.html.

Parikka, Jussi. *A Geology of Media.* U of Minnesota P, 2015.

Rancière, Jacques. *The Politics of Aesthetics: The Distribution of the Sensible.* Continuum, 2004.

Riofrancos, Thea. *Resource Radicals: From Petro-Nationalism to Post-Extractivism in Ecuador.* Duke UP, 2020.

Rosales, Antulio. "Radical Rentierism: Gold Mining, Cryptocurrency, and Commodity Collateralization in Venezuela." *Review of International Political Economy,* vol. 26, no. 6, 2019, pp. 1311–32. https://doi.org/10.1080/09692290.2019.1625422.

Selgas, Gianfranco. "Testimonios del oro negro: petróleo y memoria en el cine documental de la Venezuela del siglo XX." *Bulletin of Hispanic Studies,* vol. 97, no. 9, 2020, pp. 1003–20. https://doi.org/10.3828/bhs.2020.57.

SOS Orinoco. "Characterization and Analysis of Some Key Socio-Environmental Variables in the Orinoco Mining Arc." SOS Orinoco, 2021, https://sosorinoco.org/en/reports/characterization-and-analysis-of-some-key-socio-environmental-variables-in-the-orinoco-mining-arc/.

SOS Orinoco. "Illegal Mining, Guerrillas & Disease in the Upper Orinoco-Casiquiare Biosphere Reserve 2019." SOS Orinoco, 2019, https://sosorinoco.org/en/reports/third-report-illegal-mining-guerrillas-disease-in-the-upper-orinoco-casiquiare-biosphere-reserve/.

SOS Orinoco. "Presence, Activity and Influence of Organized Armed Groups in Mining Operations South of the Orinoco River." SOS Orinoco, 2022, https://sosorinoco.org/en/reports/presence-activity-and-influence-of-organized-armed-groups-in-mining-operations-south-of-the-orinoco-river/.

Szeman, Imre, and Jeff Diamanti. "Introduction." *Energy Culture: Art and Theory on Oil and Beyond,* edited by Imre Szeman and Jeff Diamanti, West Virginia UP, 2019, pp. 1–19.

Terán Mantovani, Emiliano. "Las nuevas fronteras de las commodities en Venezuela: extractivismo, crisis histórica y disputas territoriales." *Ciencia Política,* vol. 11, no. 21, 2016, pp. 251–85, https://revistas.unal.edu.co/index.php/cienciapol/article/view/60296.

Yusoff, Kathryn. "Geosocial Strata." *Theory, Culture & Society,* vol. 34, no. 2–3, 2017, pp. 105–27, https://doi.org/10.1177/0263276416688543.

6

Fiction Writing and Environmental Conservation in Latin America

The Cases of João Guimarães Rosa and Alejo Carpentier

VICTORIA SARAMAGO

What is the relationship between fiction writing and the creation of protected areas in Latin America?[1] This chapter explores the cases of two novels that have played a role in shaping views of regions that would later host national parks. I contend that, in spite of their remoteness, these areas were perceived as precious partially due to the symbolic importance attached to them as sites where the plots of celebrated novels unfold. As such, they are effective cases of fictional works that have helped inspire conservationist initiatives and, conversely, they make visible the ways through which the very conception of a protected area relies on a set of fictionalizing operations. The first novel is *Grande sertão: Veredas* (1956; *The Devil to Pay in the Backlands*, 1963), by the Brazilian João Guimarães Rosa, which is set in the *sertões* or backlands of the northwestern portion of the Brazilian state of Minas Gerais and neighboring states. The second one is *Los pasos perdidos* (1953; *The Lost Steps*, 1956), by the Cuban Alejo Carpentier, set largely on the region of the Venezuelan Amazon known as Gran Sabana.

National parks were created to preserve certain environments due to either their beauty or the need to address the degradation of vulnerable areas. They are par excellence spaces without human inhabitants. Since the world's first national park, Yellowstone, was created in 1872 in the United States, there has been a constant tension between parks as wilderness areas to be kept free of human presence and as institutions that serve the nation through science and tourism. The absence of human inhabitants has been at the root of some contradictions noted in classic scholarship on national parks. Roderick Nash noted how the United States' democracy has fostered the creation of national parks in their current form, but failed to note how this democracy seems to include the middle-class tourists who visit the parks, while excluding local, often Native American populations, who are expelled from them.[2] Such a view of nature obscures what Marcus Colchester has called the "unhappy truth which conservationists have only recently come to admit, [which] is that the establishment of most national parks and protected areas has had negative effects on their prior inhabitants" (103).

In spite of the environmental impact of tourism, after all, national parks usually imply a concept of nature as a pure, Romantic wilderness "out there," untamed by human hands, that does not fit well with the reality of people living and working in these spaces. National parks, thus, are meant to preserve or re-create a nature untouched by humans, at the same time as they are used for the enjoyment of humans—especially white humans. As Alison Byerly shows, the paradox is clear, since an area that receives visitors will necessarily no longer be devoid of humans. The Grande Sertão Veredas National Park, as the following section demonstrates, fully embodies these contradictions.

João Guimarães Rosa and the Grande Sertão Veredas National Park

Guimarães Rosa's only novel, *Grande sertão: Veredas,* is one of the longest books in the Brazilian canon. Without chapters, fragments, or subdivisions, it develops in one narrative block, divided only into paragraphs, as the farmer and former bandit Riobaldo recounts his life history to an educated man from a big city. The numerous *veredas* of Riobaldo's life detail his childhood memories as an orphan living with his wealthy godfather; his wanderings across the *sertão* with a group of *jagunços* (bandits), first led by the strong leader Joca Ramiro and, eventually, by Riobaldo himself; his struggles to avenge the murder of Joca Ramiro; an unconfirmed pact he might have made with the devil; and, most importantly, his conflicted love for his friend and coconspirator Diadorim in a context of absolute hostility to homoerotic desire. Echoing this narrative profusion is a geographic profusion. Alone or with his group, Riobaldo's constant travels through the region shape a dazzling geographical web that spreads into three states and encompasses hundreds of toponyms, a complex hydric system, and a number of named plants and animals. Through his blend of existing toponyms with other named geographical markers whose veracity cannot be confirmed, the geography of *Grande sertão: Veredas* draws the reader into a labyrinth: "Do you know what it is like to traverse the endless sertão, awaking each morning in a different place? There is nothing to which to accustom the eyes, all substance dissolves" (*The Devil to Pay in the Backlands* 261).[3]

I propose that *Grande sertão: Veredas* has been operating as an environmental agent in the region where the novel is set, in the northwestern sertão of Minas Gerais, in Brazil, over the past decades. As such, this work constitutes a paradigmatic example of how fiction can set standards for a reality perceived as lost in an area of accelerating environmental change such as this one. This novel's agency is not separable from the prestige of Guimarães Rosa's work in Brazil and abroad, nor can it be understood without attention to the author's intervention in the complex and multifaceted meaning the term *sertão* has acquired in Brazil. In other words, what many critics consider the "universality" of his fiction has played a real role in the conviction that its referential area of northwestern Minas Gerais should be preserved. These efforts to keep Guimarães Rosa's sertões alive have led, I argue, to a reversed form of mimesis, in which particular interventions in current realities aim to imitate fictional worlds. To talk about the sertão of Minas Gerais as a fictional environment, in this case, means to consider the expectations about this sertão based on its representation in Guimarães Rosa's works, as well as in a whole repertoire of cultural products on the many Brazilian sertões. I propose that all these implications of the term *sertão* dovetail with sustainability and environmental management projects that use the works of Guimarães Rosa as a symbolic justification for addressing the effects of deforestation and urbanization.

A unique example of *Grande sertão: Veredas*'s environmental agency is the national park named after the novel. The Grande Sertão Veredas National Park was created in 1989, amid the country's redemocratization and only a few months after Brazil's current constitution went into effect. While a number of national parks to protect the *cerrado,* where Brasília is located, were established during the new capital's construction and first year, most of the cerrado's national parks were founded during redemocratization in the mid-1980s or afterward.[4] Grande Sertão Veredas is but one example. What is unique about this park is its mission to emulate the landscape of a fictional narrative, an objective not only expressed in the park's name but repeated in official documents and reports.[5]

One of the most emblematic of such documents is the national park's lengthy "Management Plan," which details the region's history and geophysical characteristics, its flora and fauna, and the park's future directions. Produced in 2003, it dates from about a year before the park's area was more than doubled in May 2004. The frequent references to Guimarães Rosa and his work in the management plan indicate the role they played in the park's inception and development. In a section titled "Description of the area of influence," the report notes that "This whole region is widely cited in the work of Guimarães Rosa, *Grande sertão: Veredas*" (*Plano de manejo* 32). A later section, "Guimarães Rosa's setting," elaborates, "The setting of the novel is north of Minas and more specifically the cerrados of the right and left banks of the São Francisco River, whose typologies are found nearly completely preserved in the National Park" (48). It is not surprising, in this case, that a description of the use of hydric resources in the area is reinforced with the phrase "as registered in the literature of Guimarães Rosa" (78).

The counterpoint to this rich historical and cultural contextualization of the region in a management plan is the displacement of human dwellers from the park area. Their removal is only the most recent episode in a series of restrictions in access to land that the local population, also known as *veredeiros*, has endured since the 1970s. Geraldo Inácio Martins identifies four moments in this process: (1) the situation in which the novel is set, where large tracts of unoccupied and ownerless land serve as pasture; (2) when these areas began to be privatized in the 1960s, in generous programs for migrants and entrepreneurs that prompted impressive expansion of monocultures; (3) the creation of the Grande Sertão Veredas National Park in 1989, when many *veredeiros* gradually left over the following decade; and (4) the resettlement of the veredeiros, most of whom received twenty-five hectares of land in a nearby municipality.

The case of the Grande Sertão Veredas, thus, has been one of progressive expansion of the park and removal of local populations, ongoing construction of infrastructure for tourism and scientific research, and, most notably for the present discussion, an increasing use of Guimarães Rosa's work to mediate the visitors' experience of the area. When the park's first trail for tourists opened in 2015, it was lined with signs about the flora and fauna of the region that included passages of *Grande sertão: Veredas* in which they are mentioned. Here the textual environment is literally inscribed onto the real one, as if to affirm a reversed form of mimesis in which the actual sertão represents the fictional one. As such, this park aims at re-creating an original landscape that in fact invokes two different, contrasting origins: a time before the area was shaped by human activity on which the concept of national park relies, and the time of Guimarães Rosa's work in the first half of the twentieth century. Caught in between these two ideals, the veredeiros and their relocation demonstrate how problematic it is to imitate a fictional landscape marked by human presence with the conceptual tool of a national park. It is not by chance that the management plan devotes significant attention to local cultures while it narrates the very process of removing them from the park's area. The novel, on the one hand, provides a legitimized model of landscape that the park's managers are to follow, and on the other it undermines the very premise of empty, untouched space on which the very idea of national park depends. That the primary aim of creating Grande Sertão Veredas National Park was to promote environmental conservation in the cerrado is undeniable. Nevertheless, the park's name and the repeated references to Guimarães Rosa in its management plan put into relief how the Rosian sertão symbolically justifies the endeavor. A two-way process is at stake here: those who have read the author's work find reality faithful to his stories, while those who have never read his books enter for the first time into his reality.

Alejo Carpentier and the Canaima National Park

Located in the Venezuelan Guayana, the Gran Sabana or great savannah is an area primarily protected by the Canaima National Park, which was created in 1962 and expanded in 1975, becoming a UNESCO World Heritage Site in 1994.[6] Home to the world's highest waterfall, Angel Falls, and the large, remarkable table rocks known as *tepuyes,* the region has been inspiring for many decades a relatively large number of cultural products, including novels, travel accounts, films, TV programs, and publicity pieces. It is also the main site of Carpentier's novel *Los pasos perdidos.*

Carpentier moved to Caracas in 1945 and took his first trip to the Venezuelan Amazon two years later, in July 1947. Fascinated, he returned in 1948.[7] He wrote about his travels in the series of newspaper articles *Visión de América* (Vision of America) and in many other essays, talks, and journalistic contributions. Many of these pieces—and especially *Los pasos perdidos*—present the Venezuelan Guayana as enclosed and self-sufficient, in opposition to an oil-fueled modernity. While Carpentier famously called the Orinoco "a materialization of time" ("Camino" 160), the novel's compartmentalization of historical periods in the spaces of the jungle also dovetails with a certain environmental stasis. In other words, the jungle that encompasses different times, from Romanticism to the Neolithic, is a jungle relatively invulnerable to change from the outside.

The environmental changes wrought by extractivist activity and technology in the Orinoco River Basin, as Charlotte Rogers discusses in detail, alarmed the author and are visible in the novel through, for example, the presence of an airplane as well as mentions of mining and oil drilling. The contrast between such elements and Carpentier's frequent characterization of the jungle as an immutable reality akin to that of Genesis dramatizes the multifaceted, at times ambiguous set of assumptions and desires upon which his Gran Sabana is built, both in the novel as well as in newspaper articles, such as in the series that compose *Visión de América:* "Time had stopped, at the foot of the immutable rocks, devoid of all ontological sense for the frenetic Western man. . . . It was not time as measured by our clocks nor our calendars. It was the time of the Great Savannah. It was the time of the earth in the days of Genesis" ("Visión" 189). After all, Carpentier's years in Venezuela and his writings on the Amazon coincide with the country's early steps toward conservation initiatives, initially led by the Swiss naturalist Henri Pittier and culminating in the creation of its first national park.[8]

In 1953, Alejo Carpentier revisited the *novelas de la selva* genre by taking his unnamed protagonist to a compact, independent, forested world where, as a result of the area's often unpredictable dynamics, cities and other human settlements might just as easily disappear as grow.[9] As "novels of the jungle" do, *Los pasos perdidos* draws on the way the jungle fascinates and changes human characters. Carpentier tells a story of a Latin American man living in a metropolis in the Northern Hemisphere, who, after traveling to the depths of the tropical jungle, supposedly to collect Indigenous musical instruments for his former mentor's ethnographic collection, tries to leave everything behind to live and compose music there but is prevented from doing so by circumstances both from the metropolis and in the forest. Throughout this narrative, Carpentier provides profoundly rich descriptions of the mysterious, mesmerizing world of the jungle: "Despite the vast area of the jungle—he explained to me—embracing mountains, abysses, treasures, nomad peoples, the remains of lost civilizations, it was, nevertheless, a world compact, complete, which fed its own fauna and its men, shaped its own clouds, assembled its meteors, brought on its rains. A hidden nation, a map in code, a vast vegetable kingdom with few entrances" (*The Lost Steps* 126).

The references to immensity, the "vegetable kingdom" with its mountains, all correspond to a Romantic notion of wilderness quite common in nineteenth-century accounts of European travelers in America. This Romantic allusion is underscored by Carpentier's mentions of European travelers in *Visión de América,* most notably the Schomburgk brothers and Alexander von Humboldt.[10] Such isolation, however, does not preclude past and present human occupation, even though both the "remains of lost civilizations" and "wandering peoples" who are presented in a stylized fashion clearly contrast with the reality of the Pemon and other Indigenous groups actually living in the area at that time and today. This same contrast can be identified in the analogical web through which parts of the jungle materialize particular periods of world history or, in the words of González Echevarría, "a compendium of world history read backwards" (160). In other words, the deeper the narrator goes into the forest, the farther back he goes in time until he reaches "the world that existed before man" (*The Lost Steps* 186). Such representations reinforce the view of the forest as a microcosm easily detachable from the rest of the country and, indeed, the world. His slow progress by boat is quickly reversed on his way back to the country's capital, after he boards the helicopter sent to the jungle to rescue him: "One hundred and eighty minutes, returning to the epoch some identify with the *present*—as though this were not the *present,* too—flying over cities that today, at this very time, belonged to the Middle Ages, the Conquest, the Colony, or the Romantic era" (234; Carpentier's emphasis).

As a spatialization of time, a key word for understanding the forest's functioning and organization is "conservation," but not exactly "conservation" in the sense that the environmentalist movement later used the word, as Carpentier was not advocating an environmental agenda when he published this novel. Rather, the narrative conservation I refer to here has two elements. The first is the preservation of a collective environmental memory through descriptions of the forest. The second is the representation of the forest as a self-sufficient microcosm that conserves spatialized layers of world history while allowing for a dynamic of change that most outsiders do not notice. While deforestation is hardly present in Carpentier's vocabulary as either a novelist or a journalist, the narrative conservation of the jungle in *Los pasos perdidos* coincides with the development of a history of environmental conservation in Venezuela.

Unlike *Grande sertão: Veredas* and its park, *Los pasos perdidos* did not directly inspire the creation of Canaima National Park.[11] Official documents about the park do not mention the novel, and its influence may be perceived only indirectly, especially in areas with notable landmarks not mentioned in the book, such as Angel Falls and Mount Roraima. Yet the numerous descriptions analyzed above of the forest, either within or outside of the Gran Sabana area, as marvelous and uncanny may well work as carriers of a specific view of the region that the novel both participates in and propagates. More than that, *Los pasos perdidos* and essays by Carpentier clearly place the forest in a separate realm, where self-sufficiency is coupled with an apparent stasis that only obscures a series of cycles and transformations hardly perceptible to outsiders. Not only is the jungle mimetically conserved in the novel, but there seems to be a fairly clear border dividing its domain from the rest of the world.

Nonetheless, the appearance of stillness does not mean an actual pause. Carpentier's jungle actually finds itself in perpetual motion, which often reveals a long history of exchanges and mutual interactions among human and nonhuman actors. These changes include the destruction as well as the creation of cities, in a deviation from the common understanding of deforestation and urbanization as unilinear phenomena. Moreover, their manifest randomness and relative indifference to the purposes of outsiders become evident when the narrator, after having gone back to New York, is prevented from returning to the forest by a flood that has covered the marks in a tree trunk signaling the entrance to Santa Mónica de los Venados.

Taking all these factors into consideration, I contend that Carpentier's writings in his Venezuelan years, especially *Los pasos perdidos*, helped shape and were shaped by an environmental imaginary of the Gran Sabana. More specifically, there is a parallel between the history of environmental conservation in the Venezuelan Amazon and the way *Los pasos perdidos* "conserves" that forested area, both through its representation of the forest as a static, autonomous realm and through its contact points with conservationist initiatives. If the Gran Sabana's transformation into a national park and later into a UNESCO World Heritage Site is also the result of its recognizable cultural status as a precious "lost world," I propose that fiction plays a role in portraying certain areas as recognizable sites of memory that must be conserved in real life through the creation of national parks and other conservation units. While this is certainly a much more indirect process than that via which João Guimarães Rosa's work informs Grande Sertão Veredas National Park, what is at stake here is a larger circuit of many actors, of which Carpentier is but one—albeit a nonnegligible one. Yet there is a certain "structure of feeling," in Raymond Williams's terminology, underlying the forms of engagement with these areas through cultural products (*Marxism and Literature*).[12] Moreover, the very way these areas are managed may be, I argue, conditioned by certain fictionalizing operations implied in the concept of conserved areas itself—which these novels both support and critique.

Fictionalization and Conservation

Let us now put these two works together in order to articulate some theses on the relationship between fiction and conservation in Latin America. National parks try to re-create environments as they would have looked before deforestation, and their creation is premised on the (generally false) presupposition that such places were uninhabited by humans. This is the case even in parks such as Canaima, which allows Pemon villages to persist inside the park's boundaries, allowing, according to a 2000 estimate, "the approximate presence of 95 indigenous communities and possibly more than 8,000 inhabitants" (Díaz Martín and Novo Torres 11).[13] At the same time, most parks also allow non-Indigenous people to enter for scientific research and tourism. Therefore, national parks aim to re-create the appearance of an untouched nature and, in doing so, to recover an environmental memory of a certain area before the arrival of non-Indigenous humans. Even when there are Indigenous groups in the area, they are often seen through an idealized gaze that, as Luis Angosto Ferrández shows, insists on ignoring their gradual inclusion in capitalist economies and instead relegates them to an ancient temporality, in what Johannes Fabian has called a "denial of coevalness" (21).

This reconstruction, of course, would not be possible without the careful work of environmental engineering that manages and maintains this supposedly untainted nature. An example can be found in Canaima's 2007 Final Plan, where the difficulty of accessing certain areas is presented as a problem that the park's maintenance department needs to address. If the plan were really to keep it a "lost world," wouldn't this difficulty of access in fact qualify it as an "untainted" area? National parks, in this sense, constitute an effort to "civilize nature," as proposed by Bernhard Gissibl, Sabine Höhler, and Patrick Kupper: "This established notion of civilization as human progress through growth-oriented development at the cost of nature was countered by a conservationist view of civilization as an idealist endeavor that respected the aesthetic, ecological and social value of wild nature. Protecting and valuing nature in its raw, untransformed state became a property of being civilized" (9).

I argue that this maintenance of an "untransformed" appearance entails a fictionalization of the protected area through which a certain environmental imaginary is applied to it and becomes its aim. Thomas Pavel proposes understanding fictions as secondary universes that integrate a dual structure, whose base or primary universe is the real world.[14] These secondary universes are understood as possible worlds in the terms of modal logic, that is, as alternative states of affairs that maintain a certain relation with other possible worlds, including the real one or "base" (51). Religion is a good example of this dual structure and shows that such possible worlds are not necessarily understood as fictions. Fictional texts also work, although differently, as salient worlds that may include components that do not exist in the real world or that shed light on some aspect of it that is not usually perceived.

The question to ask in the case of a national park in a remote area like Canaima—and it was even more inaccessible when the park was first created—is the degree to which the secondary, fictional universe might shape the very notion of the primary universe.[15] In other words, to what extent is the area upon which the imaginary boundaries of the park are to be set seen as "raw nature" precisely because of the continuous portrayals of this area in cultural products *as* "raw nature"? In the absence of access to a primary, existent world, the secondary, fictional world takes precedence in determining how the primary universe looks and should continue to look. If the idea of establishing a protected area already implies a certain view of nature to which referential reality has to conform, this process becomes even more dependent on external perspectives in the case of an area both hard to access and frequently portrayed in cultural objects, like the Gran Sabana and the Rosian sertão. Therefore, the antithetical combination of cultural attention and difficulty of access puts this salient, fictional world of *Grande sertão: Veredas* and *Los pasos perdidos* in a privileged position to intervene in the expected state of affairs in the real world.

Conclusion

These analyses of two novels and two national parks show how narrative conservation and conservationist narratives complement each other. Narrative conservation encompasses the representation of environments in fictional works. Narratives such as *Los pasos perdidos* and *Grande sertão: Veredas* thus become carriers of environmental memories that remain active through generations and may reverberate in other cultural objects and projects. All the cultural objects about a certain environment compose a set of lived experiences that advocates for keeping it as it is or re-creating it as it once was.

In their turn, conservationist narratives are the fictionalizing operations by which parks come to be seen as "raw, untransformed" nature, even when this pristine state is only achieved through comprehensive interference in the area, whether through the removal of local populations and destruction of settlements, fire management, animal and plant control, or trail and road construction. These two modalities of interpenetration between conservation and fictionality mutually benefit each other. Fictional narratives may raise environmental awareness and promote conservationist initiatives, while national parks and other conservation units fuel the environmental imaginaries of these areas. At the same time, *Los pasos perdidos*, *Grande sertão: Veredas*, and other novels may introduce critical parameters that allow readers to perceive, for example, a longer history of human and nonhuman interaction, rather than assuming that these environments are simply "untouched."

By making this argument about the relationship between national parks and fictions, I surely do not mean to imply that national parks are fictions, or that they have any undesirable artificiality that should be rejected, or that they lack any form of legitimacy. To be sure, cultural objects do not have the same kinds of effects as does the actual maintenance of parks, which is a fundamental form of twentieth- and twenty-first-century environmental policy. Fictional works do not have the power to establish and alter policies and management practices, nor do they intervene directly in the referential reality of the park.

Yet fictional works may help in developing a set of fictionalizing operations that justify the objectives of parks, legitimize their existence, and make them more visible. This is especially true in a park of literary inspiration such as, again, Grande Sertão Veredas, but a similar conservationist mentality not seldom based on fictionalized images of pristine environments is at work in virtually any park. In fact, even parks that do not have cultural objects specifically attached to them, that are not portrayed in books, films, or TV series, and that may even be known only to a small portion of the local populations also benefit from this broader environmental imaginary that fuels a longing for untouched nature to exist.

Notes

1 This chapter originally appeared as excerpts from "The Sertão Reconstructed: João Guimarães Rosa's *Grande Sertão: Veredas*" and "Narrative Conservation and Conservationist Narratives: Alejo Carpentier's *Gran Sabana*," in Victoria Saramago, *Fictional Environments: Mimesis, Deforestation, and Development in Latin America* (Northwestern UP, 2021). Copyright © 2021 by Northwestern University Press. Published 2021. All rights reserved.
2 A variation of this process, as William Cronon notes, has taken place in the tropical rainforests since the 1970s: "Protecting the rainforest in the eyes of First World environmentalists all too often means protecting it from the people who live there" (82).

3 An early attempt at mapping *Grande Sertão: Veredas*'s toponyms was made in 1974, when Alan Viggiano proposed a bold and contested thesis: "Guimarães Rosa did not invent a single name, out of all the toponyms used in the saga of Riobaldo Tatarana" (21). Willi Bolle ("Grande sertão") revisits the issue in a detailed study of the novel's cartography that, nevertheless, remains aware of and highlights its labyrinthine dimension. Bolle shows how Guimarães Rosa's fictional geography makes slight changes in the area's real geography for narrative purposes. Unless otherwise noted, all translations were made by Laura Colaneri.
4 For an overview of the results of conservation policies in the cerrado, see Roseli Senna Ganem et al.
5 This mission also informs some of the touristic activities in the area surrounding the park. For example, a local guide well known in the region, Elson Barbosa, knows by heart long passages from Guimarães Rosa's work, which he recites to both tourists and local communities.
6 The Gran Sabana is actually a small portion of the region to which they refer. It is a specific savannah area in the broader range of *tepuyes* known as the Guiana Highlands, the highest portion of the Guiana Shield. The Venezuelan portion of the shield spreads across two states, Bolívar and Amazonas, and belongs to a single Guianan Administrative Region for the purposes of the environmental policies of the National Institute of Parks. See Isabel Novo Torres et al. Canaima National Park's original area included the Angel Falls region, today known as the Western Section. In 1975, it incorporated the Gran Sabana proper, including Mount Roraima, on the border with Brazil and Guayana, which now belongs to the Eastern Section.
7 For a more detailed account of these trips, see Roberto González Echevarría (170–74).
8 Pittier moved to Venezuela in 1917 and remained there until his death in 1950. As a result of his "intense conservationist movement" (Gabaldón 11–13), which eventually received the support of then-president Eleazar López Contreras, the first Venezuelan national park was created in February 1937. Originally named Rancho Grande, it was later renamed Henri Pittier.
9 Derided by Boom authors as lacking quality and being too committed to representing local and picturesque details, the *novelas de la selva*, as well as closely related genres such as the *novelas de la tierra* and other branches of regionalist fiction, have recently enjoyed renewed interest as sites for discussion, questioning, and reassessment of colonial and neocolonial tropes about American forests. See, for example, Jennifer French. For a critique of British neocolonialism and early environmental thinking in fictional works on the forest, see Lesley Wylie on how novelas de la selva rewrite and parody, through varying means, colonial discourses about American nature.
10 For the role of Richard Schomburgk's account in Carpentier's writing, see chapter 5 of Benítez Rojo, which proposes that the narrator's identification with Schomburgk and other proponents of "Western historical and cultural discourse" reinforces the European aspect of the narrator's identity over the Caribbean ones (211). Disagree-

ing with this centering of Richard Schomburgk's impact on Carpentier's writing, Acosta discusses the influence of Humboldt and Charles Darwin on his work (*Alejo en tierra firme* 57–68). Carpentier's handwritten and typed travel notes reinforce the view of the jungle as "dramatic," "immense," and "virgin," which attests to how present these tropes were in his thoughts from the very first contact he had with it.

11 Beyond referring to the deity Canaima in the Carib tradition, the park's name is also an obvious homage to Rómulo Gallegos's eponymous novel. One of the most prominent writers in Venezuela and onetime president of the country, Gallegos was political ally, in the Acción Democrática, of Rómulo Betancourt, the country's president at the time of the park's creation. For a description of the Pemon and their "other Carib neighbors (Macuxi and So'to)" (Sá 5) in relation to non-Indigenous literature produced about the area, including Gallegos's Canaima, see Lúcia Sá.

12 Williams defines such structures of feeling as social experiences "at the very edge of semantic availability" (134) that do not conform to "fixed forms and units" (130), which may indeed pose challenges to "formally held and systematic beliefs" (132).

13 See Díaz Martín and Novo for a description of Indigenous villages within the park's boundaries.

14 In Pavel's words, "I shall call salient structures those dual structures in which the primary universe does not enter into an isomorphism with the secondary universe, because the latter includes entities and states of affairs that lack a correspondent in the former" (57).

15 Of course, the notion of remoteness does not apply to local populations. Yet, it does apply to Venezuela's political class and most of its population at the time, not to mention an international audience aware of this "lost world" but unable or unwilling to get there.

Works Cited

Acosta, Leonardo. *Alejo en tierra firme: Intertextualidad y encuentros fortuitos*. Centro de Investigación y Desarrollo de la Cultura Cubana Juan Marinello, 2004.

Angosto Ferrández, Luis F. "Mundo perdido, paraíso encontrado: Lugar, identidad y producción en la Gran Sabana, Venezuela." *Revista colombiana de antropología*, vol. 49, no. 1, 2013, pp. 11–43.

Benítez Rojo, Antonio. *La isla que se repite: El Caribe y la perspectiva posmoderna*. Casiopea, 1998.

Bolle, Willi. "Grande sertão: cidades." *Revista USP*, vol. 24, 1994–95, pp. 80–93.

Bolle, Willi. *grandesertão.br*. Ed. 34, Duas Cidades, 2006.

Byerly, Alison. "The Uses of Landscape: The Picturesque Aesthetic and the National Park System." *The Ecocriticism Reader: Landmarks in Literary Ecology*, edited by Harold Fromm and Cheryll Glotfelty, U of Georgia P, 1996.

Carpentier, Alejo. *Los pasos perdidos*. 2nd ed., Losada, 2004.

Carpentier, Alejo. *The Lost Steps*. Translated by Harriet de Onís, U of Minnesota P, 2001.

Carpentier, Alejo. "Un camino de medio siglo." *Obras completas de Alejo Carpentier*, vol. 13, Siglo Veintiuno, 1990, pp. 141–66.

Carpentier, Alejo. "*Visión de América*." *Obras completas de Alejo Carpentier*, vol. 8, Siglo Veintiuno, 1985, pp. 169–207.

Colchester, Marcus. "Salvaging Nature: Indigenous Peoples and Protected Areas." *Social Change and Conservation*, edited by Krishna B. Ghimire, and Michel P. Pimbert, Earthscan, 2013, pp. 97–130.

Cronon, William. "The Trouble with Wilderness; or, Getting Back to the Wrong Nature." *Uncommon Ground: Rethinking the Human Place in Nature*, edited by William Cronon, Norton, 1995, pp. 69–90.

Díaz Martín, Diego, and Isabel Novo Torres. *Informe final de la evaluación del Parque Nacional Canaima, Venezuela, como sitio de Patrimonio Natural de la Humanidad*. INPARQUES, 2007.

French, Jennifer. *Nature, Neo-colonialism, and the Spanish American Regional Writers*. Dartmouth College P, 2005.

Gabaldón, Mario. *Parques nacionales de Venezuela*. INPARQUES, 1992.

Gissibl, Bernhard, et al., editors. *Civilizing Nature: National Parks in Global Historical Perspective*. Berghahn, 2012.

González Echevarría, Roberto. *Alejo Carpentier: The Pilgrim at Home*. U of Texas P, 1990.

Guimarães Rosa, João. *The Devil to Pay in the Backlands*. Translated by James L. Taylor and Harriet de Onís, Knopf, 1963.

Guimarães Rosa, João. *Grande sertão: Veredas*. Anniversary ed. Nova Fronteira, 2006.

Martins, Geraldo I. *As tramas da des(re)territorialização camponesa: A reinvenção do território veredeiro no entorno do Parque Nacional Grande Sertão Veredas, norte de Minas Gerais*. 2011. Federal U of Uberlândia, MA thesis.

Nash, Roderick. "The American Invention of National Parks." *American Quarterly*, vol. 22, no. 3, 1970, pp. 726–35.

Novo Torres, Isabel, ed. *Ciencia y conservación en el sistema de parques nacionales de Venezuela: Una experiencia de cooperación interinstitucional*. Altamira, 1997.

Pavel, Thomas. *Fictional Worlds*. Harvard U P, 1986.

Plano de manejo do Parque Nacional Grande Sertão Veredas. IBAMA, Funatura, 2003, www.ibmbio.gov.br. Accessed Sep 15, 2017.

Rogers, Charlotte. *Mourning El Dorado: Literature and Extractivism in the Contemporary American Tropics*. U of Virginia P, 2019.

Sá, Lúcia. *Rain Forest Literatures: Amazonian Texts and Latin American Culture*. U of Minnesota P, 2004.

Senna Ganem, Roseli, et al. "Conservation Policies and Control of Habitat Fragmentation in the Brazilian Cerrado Biome." *Ambiente e sociedade*, vol. 16, no. 3, 2013, pp. 99–118.

Viggiano, Alan. *Itinerário de Riobaldo Tatarana*. Instituto Nacional do Livro, 1974.

Williams, Raymond. *Marxism and Literature*. Oxford U P, 1977.

Wylie, Lesley. *Colonial Tropes and Postcolonial Tricks: Rewriting the Tropics in the Novela de la Selva*. Liverpool U P, 2009.

3

Disrupting the Sensible Order in Landscapes of Contamination, Toxicity, and Wildfires

7

Embodying Anthropocene Awareness in *ecogótico rioplatense*

ALLISON MACKEY

The gothic has proven to be a fruitful mode in which to reflect on, reiterate, and challenge contemporary cultural anxieties.[1] As we reap what feminist philosopher Rosi Braidotti calls "the disastrous planetary consequences of our species and the violent rule of sovereign *Anthropos*" (10), the observation of David Punter and Glennis Byron—that the gothic frequently reemerges with particular force during times of "cultural crisis" (39)—is particularly apt. In this study, I examine how literary engagements with haunting and monstrosity resonate with contemporary environmental anxieties in recent works by writers on both shores of the Río de la Plata: Mariana Enríquez from Argentina and Fernanda Trías from Uruguay. I argue that not only does horror appear to be a pertinent mode to reflect on, reiterate, and challenge Anthropocene anxieties, but it is also fruitful for decolonizing the Anthropocene by drawing attention to the spectral revenants of historical injustices in the Americas as they intersect with ongoing anthropogenic environmental degradation.

Through female protagonists who simultaneously represent and challenge the anthropo- and androcentric models that precede them, Enríquez and Trías demonstrate an ambiguous relationship with the contemporary moment.[2] Both the short story "*Bajo el agua negra*" and the novel *Mugre rosa* engage with gothic tropes and conventions in order to allow for the emergence of monstrous and not-fully-human assemblages. Readers share in the growing sense of horror and unease experienced by the protagonists as they navigate dystopian landscapes where rivers are devoid of life—or, at least, of life as they have known it. However, more than just a reflection of fear, both writers employ speculative narrative strategies that remind readers that the future is undetermined. It is in this liminal sense of undecidedness or lack of narrative closure that I locate a cautious and ambivalent kind of hopefulness in both narratives. In 1966 the critic Ángel Rama coined the term *raros* ("strange" or "weird") to refer to a marginal current of Uruguayan writers who characteristically distanced themselves from the realism of the nineteenth century. Faced with anxieties about the reality of our time, as Dolores Pruneda-Paz points out, "horror literature [is going through] a time of repositioning," since global pandemics, climate crises, and species extinctions "expand the margins of interpretation of the supposedly fantastic and the possibilities of the real." A growing number of contemporary writers in the Río de la Plata region are similarly devising changing modes and forms to represent changing times. Far from emphasizing a kind of horror that overwhelms and immobilizes readers, I suggest that these open-ended visions engage with ecologically inflected gothic tropes in ways that can potentially regenerate the literary imagination for the twenty-first century.

Since its emergence in eighteenth-century Britain, the generic conventions of gothic literature have spread to all corners of the world, taking root in various soils to become a useful vehicle for writers to reflect on and challenge their own forms of cultural malaise. In their research project on the "derivations" of gothic in Argentine literature, Goicochea and Guzmán-Conejeros note its flexibility as "a characteristic motif, a formula, a proportion or rhetorical quality, that does not coincide with a specific genre, but can cross different genres" (7). Understood as a mode rather than a genre, gothic sensibilities have been transplanted and cross-pollinated through distinct forms, media, genres, geographies, and cultures, escaping the confines of generic or cultural conventions in order to describe a quality of vision and thought. It is in this sense that I employ the term here. My intention is to illuminate the ways in which Enríquez and Trías "resort to the Gothic to portray the terrors of a recent past that lurks in the shadows of supposedly modernized society" (Ordiz), specifically by exposing the way that contemporary environmental concerns are indissolubly yoked to issues of social injustice in the region.

In literatures that engage the gothic mode, nature and the landscape can be disturbing sources of fear and terror: monstrosity, the uncanny, oppression, repression, and the transgression of limits can all be manifested not only through the protagonists, but also materially in nature, through land, air, and water. Barbas-Rhoden suggests that "questions of the use of land, water and resource use loom over Latin American history" (6), and the figure of the toxic river occupies a particularly disquieting role in both texts. In *Mugre rosa*, the evolution of a new species of toxic algae turns the river and the toxic "red wind" into entities of nature that pose an existential threat to human life; while in "*Bajo el agua negra*," a vengeful and uncannily bovine specter emerges from under polluted waters. I suggest that reading these narratives through an ecogothic lens can provide us with a way of identifying the ways in which these authors engage with gothic tropes in order to envision nature and landscape as spaces of crisis.

An amalgamation of ecocriticism and gothic studies, the ecogothic is a relatively new academic field (see Smith and Hughes's seminal 2013 study). As Keetley and Wynn-Sivils suggest, "adopting a specifically gothic ecocritical lens illuminates the fear, anxiety, and dread that often pervade . . . the most disturbing and unsettling aspects of our interactions with non-human ecologies" (1). It can thus be thought of as a critical movement that explores literary representations of dread or fear of the agency of that which is beyond human control or powers of domination. As Del Príncipe suggests, the ecogothic takes "a non-anthropocentric position to reconsider the role that the environment, species and non-humans play in the construction of monstrosity and fear" (1), highlighting and challenging the binaries that permeate gothic literature by showing how fear and monstrosity depend upon the transgression of these very limits.

In addition to reimagining the more-than-human space of the river, these narratives engage with the gendered material consequences of the industrial production of livestock in the region by seeking to "question the mutual oppression of women, animals and nature" (Del Príncipe 1). Further, by engaging with the ongoing environmental devastation wrought by extractivist models, both texts enter into dialogue with a subtradition of gothic literature known as "resource gothic" in order to, as Deckard points out, "figure the socio-ecological violence of extractivism, plantation and ecological imperialism in postcolonial nations" (175–76). Enríquez and Trías rectify historical amnesia surrounding the literal and figurative violence engendered by these production models by making material—"with uncanny immediacy"—"those revenants of 'undead' processes in the past that continue to shape contemporary environments" (Deckard 186). What emerges is the realization that what is in fact "monstrous" is not nature itself, but human complicity within toxic systems of production and reproduction that have led to total failures of care.

Both texts feature female protagonists caring for "monstrous" children. But both Mauro—and his unnamed medical syndrome that causes "insatiable" hunger (Trías 88), which acts as a trope to explore the pathological process in which "we consume until we destroy ourselves" (cited in Osorio)—and Emanuel, with his desire to acquire "things, sneakers and an iPhone and everything he saw on television" (Enríquez, *Las cosas* 158), are merely reflections of their societies: the logical end point or embodied consequences of capitalist accumulation. Even though they center on the figure of the human child in need of care, both texts also invite us to inhabit more-than-human perspectives and sit with the idea of the nonhuman world as also worthy of care. These narratives challenge essentialist maternal mandates, signaling "making kin" as "something other/more than entities tied by ancestry or genealogy" (Haraway 161). The "promise of monsters," to borrow an earlier figure from Donna Haraway, is tied to how both of these narratives evoke an ethics of more-than-human relationality, however "monstrous," as the only possible future in a world that has already been irrevocably altered.

"*Bajo el agua negra*" is one of several stories in the collection to focus on a middle-class female protagonist who is preoccupied with caring for, or protecting, vulnerable children in the city of Buenos Aires. As the story opens, the prosecutor is interviewing one of the police officers who has been accused of "torturing young delinquents by making them 'swim' in the river" (Enríquez, *Las cosas* 162). The body of one of the boys reappears several days later, after having awakened a nonhuman agency in the polluted river. The mutated body is understood as a more-than-human revenant who has returned from the dead to stalk the living, but, in this case, he also becomes an idol for the marginalized residents of the slum or *villa*.[3] The aspect of this story that I want to focus on here is the parallel between the social and economic "disposability" of the villa's youth and the material waste of the city. This is the only story in the collection—and possibly the only one in Enríquez's body of fiction so far—that explicitly links social injustices, understood as a kind of social toxicity, with environmental pollution.

Abandoned by the State, the villa exists along the shores of "the most polluted river in the world" (164).[4] The fetid river not only exposes (in)competent authorities, but also draws attention to the irresponsible actions of all those who have been complicit in processes of deliberate negligence. As in her other stories, in this text Enríquez draws attention to the "pacts of indifference" and "social complicity" within contemporary neoliberal societies: "If there is a Latin American horror," Enríquez says, "it is the horror of inequality" ("Narrativa de terror"). In this context, it is not enough to be "young and well-intentioned" (Enríquez, *Las cosas* 159) like the community priest: ultimately, the failure of the prosecutor's individual search for justice shows that "the perverse-malignant structure does not allow for change," and that "reform is needed in the entire social system" (Pastorino 23–24).

According to Prado and Ferrante, Enríquez's social criticism occurs "by way of the insertion of elements typical of gothic and macabre imaginary transferred to the geography and imbued with popular beliefs from the peripheral areas of the city of Buenos Aires." The prosecutor remembers the priest telling her that "no one went to church" anymore, since "most of the residents were devoted to Afro-Brazilian cults," and would set up "small altars on the corners" (Enríquez, *Las cosas* 159). Later on, the prosecutor witnesses a procession of residents swaying to the rhythm of the drums, and at the center of this strange procession is the living-dead body of Emanuel, the boy who had disappeared underwater. Many of Enriquez's stories find ways to recuperate and gesture toward local "popular beliefs and idols, witchcraft and superstitions" that, as Ramella suggests, have been rendered invisible "by instrumental and phallo-logocentric reason," and the prominence of these elements in Enríquez's work therefore helps to destabilize this "phallo-logocentric regime of meaning and representation" (136). It is interesting to consider how, in this context, the priest, as well as the abandoned church that he presides over, no longer function as central in the community and represent a hegemonic culture that has become, for all intents and purposes, irrelevant. A kind of zombie in the broader sense of the term taken by Lauro and Embry in their "A Zombie Manifesto" (2008), it is the "simultaneous occupation of a body that is both living and dead" (90). In this case, the church and the priest remain even though their function has become disconnected from the social realities of their place and time.[5] "Charged with a dark hopelessness" (Enríquez, *Las cosas* 159) and clearly out of touch with the residents of the parish for which he is ostensibly responsible, the priest grabs the prosecutor's weapon (169–70). But, before turning the gun on himself, he incongruously returns to the subject of the contaminated river, suggesting that the accumulated contamination—far from being evidence of the slow violence of a socioenvironmental crime—is what has been protecting the status quo, at least until "that boy woke up" that which was "sleeping" under "layers and layers of filth" (170).

Enriquez has explicitly said that "*Bajo el agua negra*" is her own "mixed bag" tribute to Lovecraft and his myths (cited in Navarro). Resonating with the Lovecraftian undead entity that is summoned from an underwater crypt, Emanuel "came back from the water . . . he was *always* in the water" (Enríquez, *Las cosas* 161, emphasis added). Also evoking the cephalopod Cthulhu of Lovecraft, the prosecutor notes that the girl's fingers, like those of all the "deformed" kids in the villa, are "thin, like squid tails" or like "mollusk fingers" (168, 172). Elsewhere Enríquez herself has underlined a difference between "cosmic" horror, which is based on supernatural elements or radical alterity, and terror based on "social phobic pressures" (Enríquez, "Narrativa de terror"). With the combination of supernatural and real fears that are embodied in the figure of Emanuel—neither alive nor dead, but somehow returned from the bottom of the river—Enríquez offers a critique of her own social reality: specifically, the fact that there is something in the experience "of a certain Argentine in general, which is the fear of being poor" (Enríquez "Conurbano"). This is a clear demonstration of what Mabel Moraña has called "social monstrification," a process in which "areas occupied by social sectors subalternized by dominant groups" are "*monstrified* as residual spaces whose epistemologies . . . assume unrecognizable forms from perspectives that think of themselves as epistemic, ethical and hermeneutical centers or nuclei" (13). Enriquez's homage to Lovecraft *is* a bit of a "mixed bag," in the sense that the story returns to social phobic pressures rather than insisting on any kind of cosmic Lovecraftian incommensurability. What Enríquez's narrative shows with brutal clarity is the fact that abandoning society's most vulnerable populations to a cruel neoliberal system is an incredibly short-sighted act, exposing poverty itself as a kind of monster that, sooner or later, always returns to seek revenge.

However, it is important to pay attention to the register of super/natural horror provoked in Enriquez's homage to Lovecraft, especially in relation to the emergence of the more-than-human agency of the toxic river. It may not be related to an ancient cosmic monster, but it does reflect cultural anxieties about human relationships with the natural world that are typical of the first decades of the twenty-first century. It is surely no coincidence that within the abandoned parish the prosecutor finds that in the "place of the altar there was a stick" and that "nailed to the stick, [was] a cow's head" (Enríquez, *Las cosas* 169). If Emanuel awakened a nonhuman agency that had been dormant under the water, Enríquez invites us to think of it less as a cosmic monster and more as somehow linked to the material lives of the animals that have been—like the disadvantaged youth—systematically dumped into the river:

> The houses surrounded the still, black river, skirting it and getting lost from sight where the water formed an elbow and flowed into the distance, next to the chimneys of the abandoned factories. For years there had been talk of cleaning up the *Riachuelo,* that arm of the Río de la Plata that enters the city and then moves south, that for a century has been the chosen site to dump waste of all kinds,

but, above all, the waste from cows. Every time she approached the *Riachuelo*, the prosecutor remembered the stories told to her by her father, who worked for a very short time in the slaughterhouses on the shore: how the remains of meat and bones and the dirt that the [animals] brought from the field were thrown into the water, the shit, the sticky grass. "The water turned red," he had said. "People were scared." (164)

These "abandoned factories" echo another moment in Argentina's past, specifically represented in *El matadero*, written by Esteban Echevarría between 1838 and 1840 but first published in 1871, which is widely recognized as the founding text of Argentine literature. At the same time, Enríquez's story also enters into dialogue with other recent Argentine writers, such as Fernanda García Lao, Agustina Bazterrica, and Valeria Meiller, each of whom explore the very close link not only between meat-eating and founding acts of political-religious violence, but also between meat-eating and sexual violence (see Skinner; Rossetti).

Unlike Trías, who has explicitly expressed concern for environmental issues, Enríquez has not shown a particularly ecocentric awareness in her work. However, putting her short story into dialogue with *Mugre rosa* can help us to recognize an environmental commitment that may be more ambivalent or oblique, but is no less unsettling. For Enríquez, horror "continues to be so popular because it helps us to be better prepared for the real fears . . . in our daily lives" (quoted in Navarro). Both authors have recognized unconscious influences of what Raymond Williams in the 1970s famously termed the "structures of feeling" of their place and time on their writing (e.g., see Trías and Medina; Enríquez, "The Dark"). I follow Goicochea and Guzmán-Conejeros (2016) when they suggest that structures of feeling "is a powerful explanatory concept to understand the taste for writing and reading the gothic not only in England but also in the Río de la Plata, in very different spaces and times" (6). Speculative genres—such as science fiction, fantasy, and horror—are relegated, as Amitav Ghosh (2016) notes, to the outhouses, or margins, of the literary world, but they often more clearly reflect "the tone, the drive, the heartbeat of an era" (Montes 2001). To be sure, as we enter the third decade of the twenty-first century, there are few "real fears" that are more terrifying than total environmental breakdown.

Read alongside *Mugre rosa*, the reemergence of nonhuman life at the end of "*Bajo el agua negra*" resonates less with Lovecraft and more with the "diverse earth-wide tentacular powers and forces and collected things" signaled in Haraway's notion of the Chthulucene (160). Often confused with Lovecraft's racist and xenophobic Cthulhu, the Chthulucene according to Haraway is instead etymologically connected to the adjective "chthonic," meaning of or relating to the underworld. Less about fear and anxiety, the age of earthly beings requires sim-poiesis, or doing-with—rather than auto-poiesis, or the self-creation of the Anthropos. It is about diverse human-nonhuman assemblages coming together to work for a livable future. As Haraway insists, replacing Anthropocene with Chthulucene offers us a metaphorical image that insists on the connections between all earthly beings, as a way of resisting human exceptionalism as well as committing to new ways of telling stories.

Mugre rosa is a multiple-award-winning novel by Uruguayan writer Fernanda Trías. It is set in a "possible Montevideo" (Trías and Medina), placing it squarely within the genre of dystopian speculative fiction. However, I suggest that the novel also resonates with and relies upon gothic tropes, recalling Longueil's use of the phrase "gothic imagination" to primarily describe a quality of vision and thought: or to, in Trías's own words, "recover the poetic in horror" (cited in Tanzi). Set in a run-down port city, the first-person voice of an unnamed narrator places us inside this "infected hole of old architecture" with its abandoned buildings, bricked up windows and "neo-Gothic church" (Trías 248). *Mugre rosa* can be linked to the gothic mode in its use of tropes such as enclosure in space and time, the monstrous, the uncanny figure of the double, the presence of fog, and the oppressive fear of contagion by an "enemy [that] was invisible" (16). There is a haunting déjà vu to the novel that is proper to gothic sensibility, a sense of the strange, of the uncanny, of something that "is not revealed to us completely" (cited in "Conversatorio"). This is revealed in the very structure of the text, in the disembodied and decontextualized fragments of dialogue that float between chapters like disembodied voices, as phantasmagorical as the oppressive climate of the novel.

Contamination, climate change, and the extinction of species have become part of the protagonist's daily landscape, instigating the propagation of a new species of toxic algae. The narrator notes that the occupants of what was once a hotel have left the neon sign on in order to feel that they "could still modify the landscape" (Trías 13). This particular Anthropocene awareness of human futility interweaves with the human compulsion to cross epistemological and ontological limits, something that is foreshadowed in the novel's epigraphs, and proves to be a central concern throughout.[6] A sustained doubling of the personal (affective relations and individual memory) and the public (environmental relations and cultural memory) with respect to the suspended time of mourning is complemented by a parallel between toxic codependent romantic and filial relationships, and the monstrously alienated codependencies that we have cultivated with other species: in both cases, it is a relationship that is expressed as a compulsion to consume the other. From the toxic relationships that the narrator has with her mother and her ex-husband—"he always wanted to devour me in some way" (52)—to her relationship with the "dinosaur boy" (262), to the carnivorous *pacu* fish, to the new meat-processing factory, the act of consuming the other becomes a kind of metaphor through which to diagnose a cultural disease.

It is in this sense that the novel questions the limits of monstrosity: Mauro is repeatedly called a "monster child" (Trías 258) by everyone around him (77, 182, 60, 61). Yet, charged with the task of caring for him, the narrator is well aware that it is not really the child himself but his disease that is the true "monster" (197). The novel invites us to question Mauro's monstrousness, especially in moments that demonstrate his normal childlike behavior (playing with his Lego® blocks, drawing pictures, giving kisses and hugs, as well as throwing tantrums). Juxtaposed with the behavior of the other children in the novel (e.g., the casual cruelty of a group of children who bury a poor starving dog alive), these scenes demonstrate the monstrous potential of *all* children. Above all, the image of baby Mauro's position as "a future monster, incapable of being satisfied" (71) is, in a sense, the future of all babies who eventually grow up to become adults, driven by a constant and insatiable desire to consume. To be sure, the question posed by the narrator—"what would it be like to feel constant hunger?" (71)—finds its answer on a societal level in the dystopian vision of the novel itself.

Biological maternity is also linked to monstrosity in the novel. In what Trías herself has described as a kind of mirror game, Mauro's mother is a sinister double for the narrator's mother, while the narrator herself repeats the role of her own childhood nanny, Delfa, by taking care of Mauro. The narrator and her nanny both find themselves in the position of caring for children who have been in a sense abandoned by their own mothers. But possibly the most monstrous mother figured in the novel is the new meat-processing factory: "a good mother: a provider" (113).

Reminiscent of the insidious role of language and euphemism in Bazterrica's *Cadaver exquisito*, *Mugre rosa* critiques the use of technoscientific terminologies that work to manipulate public opinion and distance consumers from the fact that this "product" was once a living being. The blurring of the line between those who eat and those who are eaten invites us to question the limits of empathy, fracturing "the taboo divisions between carnivorism and cannibalism" and imposing "the sobering reality that 'we are all meat'" (Del

Príncipe 5). This is made explicit in a particularly grotesque way in Bazterrica's novel, and somewhat more obliquely—but no less effectively—in *Mugre rosa*. Trías invites us to consider the monstrous nature of intensive livestock farming and processing through an act of imaginative empathy: after the fire that destroys the new factory, the narrator "thought about the animals: no one would have considered them victims" (196). In turn, Trías places the woman worker who produces chicken nuggets in the factory in the same dehumanized and mechanical position as the animals she processes (189).

Mugre rosa denounces a fundamental failure of care by exposing how the people who should be taking care of us are in fact not doing so. This is figured not only through the narrator's and Mauro's failed mothers, but also the mysterious destruction of the "mother" factory, as well as the suicidal mother who jumps from the top of a building with her child in her arms (Trías 25), and the idea of vengeful "mother nature" acting through a mutated, killer algae. Fears related to the "deep indifference" (190) of nature toward human lives can be understood as ecophobic, to be sure, and the novel exhibits a melancholy sense of ecoanxiety that is typical of Anthropocene structures of feeling: "Who could have imagined the auditory lack in a city without insects, without buzzing, but also without horns, without the slow snort of an elevator or the murmur of distant voices, without all those artificial sounds that—I now understand—was what we called life" (213). From the sudden spectacle of dead fish on the beach to the slow disappearance of the birds that "left *us* alone with the red wind" (95, my emphasis), the novel reflects a sense of what Albrecht et al. have called "solastalgia," the existential anguish caused by negatively perceived environmental change. The novel is narrated in the past from an unspecified future moment, reflecting on end times from the perspective of an exclusively anthropocentric "us."

At the same time, however, the novel's temporal suspension in the present tense helps Trías's vision move beyond ecophobia. *Mugre rosa* is positioned somewhere between the past that has led us to this moment and the not yet of a dystopian future, and this is especially showcased in Trías's innovative structural experimentation. This liminality is especially indicated in the figure of the narrator herself, who is suspended in a time of both personal and environmental mourning, frozen between a circumscribed past and an open future: "The moment would come," says the narrator, when "there *would be* no animal in the sea that was not a mutation" (235, my emphasis). The curious use of the conditional future here—"would be"—when the rest of the narrative is written in the past or the simple future tense reminds us that the future is yet unwritten. There is a window of possibility that things might still be otherwise. The speculative nature of Trías's novel demonstrates the ethical and aesthetic potential of narratives that simultaneously look to the past as well as the future: writing speculative fiction is, as the writer Lidia Yuknavitch suggests, a question of infiltrating the present tense through imagination to shake the state of normality, and this is where a cautious and ambivalent kind of hopefulness lies in both narratives.

Both protagonists are simultaneously beholden to past generations and harbingers of possible futures. They try to free themselves from the attitudes of the past, something that is appropriately figured in a heavy backpack full of (unread) books that the narrator carries in *Mugre rosa* (Trías 21), and attempts to liberate from familial legacies resonate the typically gothic trope of parricide (although in the case of *Mugre rosa* it is, properly speaking, a kind of figurative matricide). In "*Bajo el agua negra*," the prosecutor remembers her father's "pompous, chemistry professor air" as he gives her scientific explanations of the "anoxia" of the lifeless river (Enríquez, *Las cosas* 164), which is reminiscent of the euphemistic technoscientific language in *Mugre rosa* that resonates with the rationalist logic of contemporary production, where "everything that is inconvenient has a technical name, insipid, colorless, and odorless" (Trías 49). Both the chemist father and the agronomist represent previous generations that have been responsible for increasingly efficient (and destructive) models of production.

In the gothic tradition, miasmic mist and fog are often signs that there has been a blurring of boundaries between good and evil, but this novel offers an important twist that reflects the climatological conditions of the Río de la Plata region: instead of being associated with nefarious or mysterious forces, the fog—and the *lack* of vision and clarity associated with it—actually protects and comforts the narrator. The fog offers safety from winds that bring toxicity along with "clarity" (106). The narrator expresses an anxiety that resonates particularly with the Anthropocene moment: "If something characterized confinement it was that feeling of no time. We existed in a waiting that was not really waiting for anything concrete either. We waited, expectantly. But what we hoped was that nothing would happen, because any change could mean something worse. As long as everything was still, I could maintain the no time of memory" (105). Here, the use of the plural pronoun "we" interpolates us as contemporary readers. The sense of comfort or security that comes from a *lack* of clarity is indicative of a reluctance to see beyond our current moment. The fears of dangerous nature that once stirred us—"scorpions, thistles, jellyfish, giant waves" (96)—are no longer the biggest threats to our existence. Like the first witnesses to see the algae turn the river a sinister red, even taking photos of the "beautiful" phenomenon (72), we are not prepared to understand the extent of the damage we have already done to the planet, nor do we really understand that the fear that we should feel is not *of* nature, but of the consequences of our lack of caring for it. In *Mugre rosa*, confinement goes beyond the narrator's house or even the port city itself: the home that has become strangely sinister and increasingly *unheimlich* for humans is the earth itself.

In another game of mirrors, the living "statue" of a woman who is trapped between "paralysis and inertia" (Trías 160–61) reflects the narrator's own immobility. Later, she finds another disturbing double in the "dying" and monstrous caged bird (215). Once the door to her own emotional cage opens, she, like the bird, does not flee. At the end of the novel, Mauro's mother is "expecting" (261) a future symbolized by the arrival of a new baby, as if the figure of the child could still represent futurity or provide some kind of hope. In this sense, Mauro's mother, as well as the narrator herself, act as haunting doubles for us as Anthropocene readers, since we also find it more comfortable (and comforting) *not* to see clearly what is really happening around us: overwhelmed by multiple and overlapping climate crises, the collapse of ecosystems and the extinction of species, we are similarly suspended, caught between a world that we could once depend upon and an uncertain future.[7] This moment of temporal suspension, this "no time of memory" (105) that is structurally performed in the novel, resonates with a kind of utopian impulse, but instead of "no-place" rather points toward alternative temporalities.

Gothic female characters traditionally demonstrate an excess of sensitivity and an out-of-control imagination, and in these novels, the protagonists find themselves branded as crazy or "stupid" (Enríquez, *Las Cosas* 169, 170) by previous generations. However, part of their break with the past involves embracing a break with the rationalist ideal of the clean and proper body. Unlike her ex-husband, who seeks "to separate himself from his body" (Trías 16), the narrator of *Mugre rosa* eventually accepts that she is physically connected to place as she reaffirms a kind of kinship with the polluted river: the "port, a mixture of algae and spilled gasoline. That smell is mine" (269), she says. She refuses to be relocated by the authorities, and finally moves toward the future on her own terms; with "neither direction nor plan," she realizes that she cannot "stop a future that is already here" (274). At its core this is a novel about grief, or, more specifically, about the temporal process of grieving. The fragments of text separate the chapters and thereby allow a moment of suspension—the "no time of memory" (105)—to reflect the tendency to dwell on what is already gone, instead of looking to the future or moving on. In this case, the protagonist's personal grief coincides with a collective grief for a world that has been—or is still in the process of being—transformed beyond recognition.

More than just a focus on grieving loss, however, adaptation to that which has been previously unthinkable is also central to both texts. The spectacular and catastrophic image of the "last fish" can easily distract us from noticing that, actually, "some fish adapted" (Trías 47). The "distant relatives" of the canned tuna that the narrator eats with sensual abandon are "unknown mutations" that already "swim in our rivers" (235). Similarly, the narrator admits that there is a kind of beauty in "the red moss" she finds growing in an abandoned and boarded up building (214). Like the chronically ill patients who "guard the secret of the algae" in their bodies (38), monstrous mutation might also be understood as a continuation of life beyond human-centered modes of existence, as alluded to in the narrator's fascination with stories about the adaptation and survival of nonhuman animals of Chernobyl.

With this in mind, the last image of the prosecutor crossing the bridge in Enriquez's "*Bajo el agua negra*" betrays a cautious and ambivalent sense of possibility:

> She ran, trying to ignore the fact that the black water *seemed* agitated, because it *could not* be agitated, because that water *did not* breathe, the water was dead, it *could not* kiss the shores with its waves, the surface *could not* stir up in the wind, it *could not* have those eddies or that current or have risen, *how was that possible* if the water was stagnant. Marina ran toward the bridge and didn't look back and covered her ears with her bloody hands to block out the noise of the drums. (174, emphasis mine)

Even Enriquez's use of language itself becomes unhinged in this passage: but what I find particularly interesting is the repetition of the phrase "could not." The logic that dictates what is "possible"—what is, in fact, thinkable—does not mesh with the information that the prosecutor receives with her own physical senses. The sense of possibility and openness that I find in this story is reflected in the idea that, with one foot on either side and suspended in a liminal moment that echoes the moment in Trías—that moment when, "as long as Coyote didn't look down, as long as he didn't realize that he was running in thin air, with no earth under his feet, he didn't start to fall" (229)—the prosecutor just might open her eyes, lower her hands, and unblock her ears.

In both narratives, rivers and the mutated beings that live within them function as "narrative actants" in Jane Bennett's sense. Bennett argues that dissolving the subject-object binary requires that we "begin to experience the relationship between persons and other materialities more horizontally" (10). Bennett's theory of distributive agency expresses the active powers that emanate from things, challenging the notion of "agentic capacity" as limited to human intention. Agency, she argues, needs to be "distributed across a wider range of ontological types" so that things can be recognized as having the ability to produce effects (10). In these novels, algae and fish mutate to survive in increasingly toxic aquatic environments, and an adolescent "learns to swim" while awakening the more-than-human agency of an ostensibly "dead" river: in both cases, any possible future depends on mutation or adaptation because the damage has already been done and there is no way back. The dead-but-also-somehow-still-alive rivers, and the entities that survive within them, against all thinkability, remind us of the ontological relationship of the human animal to the already toxified but resilient earth, and to the other material beings of the Chthulucene, as "assemblages" that are "living, throbbing confederations that are able to function despite the persistent presence of energies that confound them from within" (23, 24).

In Bennett's notion of "vibrant matter" nonhuman agency is made visible through the frustration of human intentionality: "By 'vitality' I mean the capacity of things—edibles, commodities, storms, metals—not only to impede or block the will and designs of humans but also to act as quasi agents or forces with trajectories, propensities, or tendencies of their own" (viii). This is the narrative power of the material world, which invites us to recognize its irruption as a kind of "storied matter" (Iovino and Oppermann), in order to foster new types of narratives that have less harmful effects in the world of embodied nature. As Bennett ponders: "How would an understanding of agency as a confederation of human and nonhuman elements alter established notions of moral responsibility and political accountability?" (21). As readers in the present moment we might realign ourselves with the ethical realignment experienced by their protagonists and recognize that a multispecies ethics of care can destabilize anthropocentric dualisms. Inhabiting an Anthropocene awareness, we can continue to deny that "something" has gone "horribly wrong" (Enríquez, *Las Cosas* 169), or we can push past the discomfort in order to face up to the horrors of which we have all been complicit.

Notes

1 This chapter is a translated, revised, and condensed version of "Aguas Ambiguas: Encarnando una Conciencia Antropocénica a través del Ecogótico Rioplatense." *Revista CS*, no. 36, 2022, pp. 247–87. All translations from Spanish to English are the author's.
2 I read these texts as Latin American examples of what Trexler identifies as "Anthropocene fictions," stories that are not only *about* the Anthropocene but also *of* the Anthropocene in the way that they employ narrative strategies to display an emerging epochal self-awareness, thus marking a limit or a liminal moment in cultural production.
3 The figure of disappeared bodies has a literary trajectory that is related to state repression in Argentina's not-so-distant past, echoing the horrors of the military dictatorship, when political prisoners—sedated and bound—were thrown alive into the Río de la Plata from airplanes. The disappearance of the boy in the river also reminds readers of more recent events in Argentine history, for example, the highly social mediatized case of Santiago Maldonado.
4 Alas, this river is not a fictional invention. Enríquez writes autobiographically about the same river in "Riachuelo" in the collection *Tales of Two Planets: Stories of Climate Change and Inequality in a Divided World*.
5 Rich in allegorical meaning, the zombie has become a universally recognized figure in popular culture. Originally a figure of Haitian folklore, representing the brutality of a life of forced labor without end, in ensuing decades there has been a shift in emphasis from the Haitian zombi as connected to slavery, to the modern slavery of the capitalist system. While I agree with Lauro and Embry that "we must not disconnect the zombie from its past" (98), in very general terms becoming "zombified" these days means succumbing to control by an outside force and losing autonomy, while continuing to resemble what (or who) it once was. It is in this sense (e.g., zombie capitalism or zombie consumerism) that I invoke the trope here, in order to describe a being or a system that is controlled by another or is mindless.
6 For example, the image of the "mass of crabs" in the sand, when the protagonist feels "for the first time . . . that there was something incomprehensible, bigger than us" (15), later reminds her of her mother: "It generated the same uneasiness, the same primitive fear" that is later expressed in explicitly Gothic terms in an image of her mother as a haunted house, "full of nooks and crannies and false doors" (26, 58).
7 This moment makes it difficult to see things clearly. As Amitav Ghosh points out in *The Great Derangement*, we cannot even adequately represent our situation as something that is *really* happening to us in the present, instead relying on the comforting "what if-ness" of dystopian speculation.

Works Cited

Albrecht, Glenn, et al., "Solastalgia: The Distress Caused by Environmental Change." *Australasian Psychiatry,* vol. 15, no. 1, 2007, pp. 95–98.

Barbas-Rhoden, Laura. *Ecological Imaginations in Latin American Fiction.* UP of Florida, 2011.

Bennett, Jane. *Vibrant Matter: A Political Ecology of Things.* Duke UP, 2009.

Braidotti, Rosi. *Posthuman Knowledge.* Polity, 2019.

Conversatorio sobre *Mugre Rosa* de Fernanda Trías con Piedad Bonett. Facultad de Artes y Humanidades de la Universidad de los Andes, 2021, https://www.youtube.com/watch?v=jAfwVDC8Q3g&t=2341s

Deckard, Sharae. "Ecogothic." *Twenty-First Century Gothic: An Edinburgh Companion,* edited by Maisha Wester and Xavier Aldana Reyes. Edinburgh UP, 2019, pp. 174–88.

Del Príncipe, David. "Introduction: The EcoGothic in the Long Nineteenth Century." *Gothic Studies,* vol. 16, no. 1., 2014, https://doi.org/10.7227/GS.16.1.1.

Enríquez, Mariana. "Conurbano: Mariana Enríquez." *Canal Encuentro,* 2017: https://www.youtube.com/watch?v=Bx__crZR02M

Enríquez, Mariana. "The Dark and the Hidden." Interview with Mariana Enriquez and Guadalupe Nettel by Peter Adolphsen, Louisiana Literature Festival, Louisiana Museum of Modern Art, 2018, https://channel.louisiana.dk/video/mariana-enriquez-guadalupe-nettel-the-dark-and-the-hidden.

Enríquez, Mariana. *Las cosas que perdimos en el fuego.* Anagrama, 2016.

Enríquez, Mariana. "Narrativa de terror por Mariana Enríquez." *FLASCO Argentina,* 2017: https://www.youtube.com/watch?v=bHdM7Wq6fe4.

Ghosh, Amitav. *The Great Derangement: Climate Change and the Unthinkable.* U of Chicago P, 2016.

Goicochea, Adriana Lía, and Rodrigo Guzmán-Conejeros. "Derivaciones del modo gótico en la narrativa argentina. Las generaciones de postdictadura." *Anuario Pilquen. Sección divulgación científica del Curza,* vol. 1, no. 1, 2016, https://revele.uncoma.edu.ar/index.php/anuariocurza/article/view/1844.

Haraway, Donna. "Anthropocene, Capitalocene, Plantationocene, Chthulucene: Making Kin." *Environmental Humanities,* vol. 6, no. 1, 2015, pp. 159–65.

Haraway, Donna. "The Promises of Monsters: A Regenerative Politics for Inappropriate/d Others." *The Monster Theory Reader,* edited by Jeffrey Andrew Weinstock. U of Minnesota P, 2003, pp. 459–521.

Iovino, Serenella, and Serpil Oppermann. *Material Ecocriticism.* Indiana UP, 2014.

Keetley, Dawn, and Matthew Wynn Sivils. "Introduction: Approaches to Ecogothic." *Ecogothic in Nineteenth Century American Literature.* Routledge, 2018.

Lauro, Sarah Juliet, and Karen Embry. "A Zombie Manifesto: The Nonhuman Condition in the Era of Advanced Capitalism." *Bound* 2.35, no. 1, 2008, pp. 85–108.

Longueil, Alfred. "The Word 'Gothic' in Eighteenth Century Criticism." *Modern Language Notes,* vol. 38, no. 8, 1923, pp. 453–60.

Montes, Graciela. "El mundo como acertijo." *Brecha*, Uruguay, May 18, 2001, https://www.lainsignia.org/2001/mayo/cul_069.htm.
Moraña, Mabel. *El monstruo como máquina de Guerra*. Iberoamericana, 2017.
Navarro, Elisa. "Las obsesiones de Mariana Enríquez." Zero grados, 2017, http://www.zgrados.com/las-obsesiones-mariana-enriquez.
Ordiz, Inés Alonso-Collada. "Civilization and Barbarism and Zombies: Argentina's Contemporary Gothic." *Latin American Gothic in Literature and Culture*, edited by Inés Ordiz and Sandra Casanova-Vizcaíno. Routledge, 2018, pp. 15–26.
Osorio, Camila. "Fernanda Trías: 'Consumimos hasta destruirnos a nosotros mismos.'" *El País Cultura*, January 6, 2021, https://elpais.com/cultura/2021-01-06/fernanda-trias-consumimos-hasta-destruirnos-a-nosotros-mismos.html.
Pastorino, Agustina. "El mal social como fuente de terror: un recorrido por la narrativa breve de Mariana Enríquez." Undergraduate thesis, Universidad de San Andrés, Buenos Aires, Argentina, 2018, https://repositorio.udesa.edu.ar/jspui/handle/10908/16623.
Prado, Esteban, and Lucio Ferrante. "Devenir americano del terror argentino: un diálogo crítico con Franco Bifo Berardi." *Recial*, vol. 11, no. 17, 2020, pp. 142–67.
Pruneda Paz, Dolores. "Literatura argentina de terror: un fenómeno que crece al ritmo de premios y traducciones." *Infobae Cultura*, September 17, 2021, https://www.infobae.com/cultura/2021/09/17/literatura-argentina-de-terror-un-fenomeno-que-crece-al-ritmo-de-premios-y-traducciones/.
Punter, David, and Glennis Byron. *The Gothic*. Blackwell, 2004.
Rama, Ángel. "Prólogo." *Aquí: Cien años de raros*. Arca, 1966.
Ramella, Juana. "El reencantamiento terrorífico del cuento argentino: Mariana Enríquez." *Boletín GEC*, vol. 23, 2019, pp. 122–38.
Rossetti, Lucía Caminada. "Rituales políticos, sexuales y sagrados en la literatura del siglo XIX. *El Matadero* como espacio de transición y mezcla." *Amérique Latine Histoire et Mémoire. Les Cahiers ALHIM*, vol. 29, 2015, doi: 10.4000/alhim.5268.
Skinner, Lee. "Carnality in 'El matadero.'" *Revista de Estudios Hispánicos*, vol. 33, no. 2, 1999, pp. 205–26.
Smith, Andrew, and William Hughes. *EcoGothic*. Manchester UP, 2013.
Tanzi, Silvana. "Yo quería recuperar lo poético en el horror." *Semanario Búsqueda*, October 28, 2020, https://www.busqueda.com.uy/Secciones/-Queria-recuperar-lo-poetico-en-el-horror--uc2079.
Trexler, Adam. *Anthropocene Fictions: The Novel in a Time of Climate Change*. U of Virginia P, 2015.
Trías, Fernanda. *Mugre rosa*. Random House, 2020.
Trías, Fernanda. Interview with Fernando Medina. "Con Fernanda Trías: oír con los ojos." *En perspectiva*, October 17, 2020, https://enperspectiva.uy/en-perspectiva-radio/oir-con-los-ojos-fernanda-trias-autora-mugre-rosa/.

8

Haunting Trees in the Global South

Image and Life in the Rubble of Climate Change

Roberto Robalinho

A report from the IPAM (Amazon Environment Research Institute), published during the Glasgow COP26 summit, states that the Amazon Forest has been, since 1990, gradually emitting more carbon than it is able to retain due to degradation and forest fires (Silva et al.). The main argument is that the fire-degraded forest is not only less efficient in storing carbon but also acts as a carbon emitter. The title of this report is "The Hidden Emissions," which highlights a key point regarding our current ecological crisis: our capability, or, rather, incapability, to see and measure the intensity of our present state of destruction. The scientific disputes regarding climate change, where objective evidence and scientific methodology play an important role, are not new.[1] However, it is important to think about how the title of the study emphasizes that there is something hidden—the invisible emissions—that must be made visible. This reveals how climate change is also a dispute over visibility within a visual regime, where an *economy of the image* operates in terms of Marie-José Mondzain, "those who are the masters of the visible are the masters of the world organising and controlling the gaze" (3). This means that fighting climate change is not only about producing scientific evidence but also about managing a visual regime that determines how we see and understand the causes and effects of climate change.

The aim of this chapter is to reflect on the relations between image and the Anthropocene by looking at specific events in the Global South. The term "Anthropocene" was proposed by Nobel Prize–winning chemist Paul J. Crutzen and collaborator Eugene F. Stoermer in 2002, and it suggests we now live in a new geological era marked by how human activity, mainly the extensive use of fossil fuels, that has altered the conditions of the geosphere.

The term has triggered broad discussions by scholars from different research fields and is not a consensus. One of the controversies related to the Anthropocene is how to define the geological markers and the beginning of this new era. There is also the problem of scale: How can we understand a force so vast that affects every living being, human and nonhuman, globally? How do we measure the effects of the Anthropocene that are everywhere but lived locally and felt in different intensities by different beings?[2] My intention is to discuss the visualities of the Anthropocene by looking at images from three different Global South events that we shall call "scenes": first, the rupture of a mining dam in Brazil in 2015; second, the destruction of a sacred ghost gum tree in Northern Australia in 2013; and, third, the Australian wildfires of 2019/20. Each of these events has its own singularities and specific histories, and the focus is not to erase these differences, but rather look at how, in their differences, they mobilize a similar aesthetic regime. In the case of the rupture of the mining dam in Brazil and the Australian wildfires, images are an important tool in framing an apocalyptic experience of a world that is falling apart. However, against this end-of-the-world scenario, photographs of the aftermath reveal hidden subjectivities neglected by a modern colonial gaze, while, at the same time, they bring to the fore the dense and entangled temporalities of the Anthropocene. The same unwanted subjectivity is also present in the 2013 destruction of the sacred ghost gum tree in Australia, where the portraits of trees have an uncanny incorporation that disarranges our constitutive modern separation between culture and nature, an experiential/visual discomfort and ontological danger that results in the material destruction of the portrayed trees.

These different and singular events are tied together by the images of haunting trees that surface in the aftermath of destruction. What is really striking is how the image is produced by the trees much before the framing of the camera. This image, which resembles a mark drawn on the landscape, has a double nature. On one level, it reveals an agency from the tree that is involved in producing this mark. On the other, it represents the tree's strength withstanding a violent catastrophe (in the Brazilian case) or the tree ghost, surviving the intensity of the Australian fires. It is only later that photographers will arrive at the scene and activate this virtuality—the tree's agency—through their photographs. Alfred Gell's theory of art and agency can help us think about this operation and superposition of images: the lines drawn on the landscape by/with the trees and the photographs taken by photographers. Although Gell is concerned with an anthropological approach to art objects, distancing from a representational and symbolic perspective, his insistence on looking at art through an "emphasis on *agency, intention, result,* and *transformation*" (7), is what interests us. Gell sees *agency* in artifacts as relational, where there is always a *patient* and *agent* relation between *artifacts,* artists, and beholders (21). These are not solid positions and are interchangeable and depend on context. What is important to understand is how the agent is defined by the ability to affect the patient and the effectiveness of his actions. The trees are agents in relation to the catastrophe when they mark on the landscape the passage of the destruction. They are also agents when the photographers arrive at the scene since the mark is not a simple mark, but one that reveals the tree's bodies and their agency in producing this image—an image that denounces the colonial destruction, but also reveals the bodies of the trees, evoking a subjectivity neglected and willfully suppressed by modern epistemology.

Visualizing the Anthropocene

The concept of the Anthropocene has been criticized for how it generalizes an idea of the human, one that is very disembodied, not gendered, racialized, or geographically and historically located (Chakrabarty 214; Latour 111; Haraway 43). For the discussion proposed here, let us locate our understanding of the Anthropocene within the idea of the Plantationocene, for it relates to an economic model that is deeply connected with the Global South and the colonial enterprise (Haraway et al. 557). Historically, the plantation system is a socioeconomic technology based on the extraction of natural resources and the exploitation of racialized bodies and subjectivities, a system that persists in the present in many ways. On the one hand, the technical ecology of the plantation is based on the efficiency of a monocrop that alienates local biodiversity and knowledges, imposing a modern view on southern natural landscapes and cultures. For the plantation enterprise to prosper, according to Macarena Gómez-Barris (5), a territory is designed and made extractive. An extractive economy not only produces a radical separation between men and nature but also a violent separation between those destined to be masters and those to be slaves. This violent commodification of nature and marginalized peoples is also what makes the intense primitive accumulation of capital possible.

Nicholas Mirzoeff calls the plantation system the first "complex" of the "complexes of visuality." It is a visual system created to support power and sovereignty in the colonies, a system that is intrinsically entangled with modern ways of seeing (10). According to Mirzoeff, there are two sides to visuality: one is the visual and surveillance technologies related to bodies, populations, nature, and the land. Techniques such as cartography, taxonomy, and anthropometry, developed in and with modernity, describe and objectify landscapes and bodies. The second side is a sensible aesthetic regime that sustains these racialized socioecological divisions and hierarchies, one that is not only produced by media representations, such as literature, journals, prints, and paintings but also through architecture, laws, philosophy, costumes, urban spaces, and so on. Visuality becomes an important conceptual frame to understand how a power regime controls the colonial space and renders it profitable.

If we consider Global South socioecologies shaped by the Plantationocene and how its long history results in the emergence of climate change, we must also consider the hidden visualities that organize and regulate this system. This means that, beyond scientific disputes, in the context of climate change, there is also a political struggle that can be seen through different countervisualities. Some of these countervisual strategies are actively produced by different social actors, such as activists, artists, and Indigenous populations. However, many images are contingent on ecological disasters, created during and after the destruction by natural elements that are uncontrollable. These images, which come from an environment affected by men, can become disruptive through a series of humans and nonhuman agents, as the scenes that shall be analyzed here.

In *The Falling Sky*, Yanomami shaman Davi Kopenawa insists that white men are unable to see with their eyes, due to a visual regime that ignores the invisible beings that live in the forest.[3] The book, written in alliance with anthropologist Bruce Albert, is an attempt to tell Kopenawa's tale and open our eyes to the spirits that live in the forest and help to preserve it:

> But white people who clear the forest probably think that its beauty came to be for no reason? It isn't true! They only ravage it without worrying because they cannot see it with shamans' eyes. In the places that they occupy, all that remains are savannas and a soil that has lost its breath of life. But as long as we live here, this will not happen! (386)

In the first part of the book, Kopenawa narrates poetically how he became a shaman, which is a process of learning how to see the xapiri, the magical spirits that inhabit the forest. He poetically tells us how he managed to understand that the bright glittering light he saw was the xapiri dancing with their mirrors. The spirits maintain the forest and need attending to in a radical cosmopolitical perspective:

> In our very old language, what the white people call "nature" is *urihi a,* the forest-land, but also its image, which can only be seen by the shamans and which we call Urihinari, the spirit of the forest. It is thanks to this image that the trees are alive (Kopenawa and Albert 389)

It is interesting to think of how the translation uses the word "image" to name this double, this force that inhabits what, according to Kopenawa, we call "nature."[4] There is a practical proposition in the book that white men must understand how the forest is seen by the Yanomami shaman so that we can understand the ontological risk of destroying and commodifying the forest.[5] According to Kopenawa, we are all connected by the same vital force that holds the world as it is, with all its diversity. To preserve the forest is to care for this force and the world itself. We can argue, from Kopenawa's words, that there is a problem with giving visibility to the xapiri that the white men are unable to see. In the core of the ecological and cosmopolitical proposition by the Yanomami shaman, there is a new visual order where the forest's hidden virtualities represented by the xapiri need to be seen. Virtuality here must be understood in Viveiros de Castro ("Perspectival Anthropology") perspectivist terms, where the forest is coextensive to Indigenous sociality and is inhabited by a multiplicity of Others—humans and nonhumans.[6]

Other proposals framed within critical perspectives of the Anthropocene also address the question of visuality and visibility. Among many different authors, there is an effort to make visible a hidden ecological destruction, but also the life forms that survive in the devasted and endangered landscapes. In Rob Nixon's concept of slow violence, based on a "delayed destruction that is dispersed across time and space" (2), the invisibility of this violence is crucial. Contrary to other spectacular forms of violence present in many ecological catastrophes, slow violence is always "out of sight," even if it is continuously and anonymously constituting a devasted landscape. As the author insists, the question of giving visibility to an invisible, structural violence is urgent:

> How can we convert into image and narrative the disasters that are slow moving and long in the making, disasters that are anonymous and that star nobody, disasters that are attritional and of indifferent interest to the sensation-driven technologies of our image-world? (3)

Scene 1: A Thick Red Line

On a late afternoon in November 2015, in the district of Mariana in Minas Gerais, Brazil, a containment dam of the Samarco Mining Company burst, releasing 50 million cubic meters of red toxic mud that immediately covered the village of Bento Rodrigues. The toxic mud continued spreading down the mountains of Mariana until it reached the Rio Doce—the sweet river. The thick mud continued downstream, and, days later, it reached the sea, at almost 650 kilometers away from Mariana. The collapse of the dam killed nineteen people and displaced seven hundred others. It exterminated wildlife and ecosystems and contaminated drinking water for hundreds of thousands of people. For the Krenak Indigenous people, the death of the Rio Doce was even more traumatic since they call themselves the Borum du Watu—the people from Rio Doce. In many ways, the Krenak "make kin" (Haraway 99) with the river, and losing it was like losing a parent, an ancestry.[7]

Samarco is a company owned by the biggest mining companies in the world. Two of its most important shareholders are Brazilian Vale and Anglo-Australian BHP Billiton. The Mariana dam disaster was the biggest environmental accident in Brazil until, in January 2019, another mining dam, operated by Vale in Brumadinho, about 120 kilometers away from Mariana, collapsed, killing 257 people. The media coverage of these tragic events was saturated with apocalyptic images of the destruction.

Five months after the Mariana disaster, photographer Bruno Veiga published the photo essay *Red Desert*.[8] The title is a reference to Italian filmmaker Michelangelo Antonioni's homonymous movie.[9] In Antonioni's film, a bourgeoisie housewife suffers from an unexplained anguish, and Antonioni uses the devasted industrial landscape of the Po Delta in Northern Italy to express the main character's anguish. Antonioni paints industrial pipes, valves, and tins with an artificial bright red paint, detaching the industrial elements from the pale green misty delta, producing a clear discontinuity between the natural landscape and the industrial plant. In an interview by the Brazilian poet and literary academic Eduardo Sterzi, Veiga said that he used Antonioni's reference because he found in the devasted landscape of the *Rio Doce* the same "ill being" he found in the film (Veiga and Sterzi).

Bruno Veiga's photographs register a perfect red line drawn by the thick mud in the landscape. According to Sterzi, these photographs evoke the complex temporality of the event, revealing at the same time the acceleration of the mud rapidly devouring the landscape, and also the deadly stillness of the aftermath of the disaster (Veiga and Sterzi). The red line freezes the acceleration in the present—an image that is there even before the photograph. The uncanny line is where the catastrophic spectacular violence meets Nixon's slow violence, colliding two temporalities of the Anthropocene: the unavoidable cataclysmic apocalypse and the ongoing silent extractive violence. What is remarkable is how, in the case of these photographs, Henri Cartier-Bresson's "decisive moment" (Hermanson 31) is not produced by the photographer whose sharp eyes and fingers are capable of freezing a moment of life.[10] Rather, it is the body of the tree, against the violence of the mud, that freezes and upholds a moment of death. Veiga displaces his body to the devasted landscape and in this movement sees the trees that brushed against the mud. His photographic act is not one of creating an image or suspending an instant, but of showing something that is already there: the hidden virtuality of the gesture of the trees that served as a barrier to the mud and now have a mark of a collision of temporalities tattooed on their bark.

The trees in Veiga's photographs also incorporate a body, a presence, that both produces this entangled temporality and is the body that withstands the catastrophe. One could say that the trees are the witnesses that survive and speak through their bodies painted by the thick mud. However, the trees are not only exposing overlapping temporalities, but they are also incorporating a body in the present. There is an evident corporality to the photos that reveals both the critical fragility of nature and the resilience that manages to exist in the breaches of the thick red mud. In this sense, the photos mediate a relation between the trees, the photographer, and the beholders of the photograph, where the image of a body emerges. The trees' agency is made visible when their body acts as a canvas for the mud's thick red line. Once the trees' marked body is photographed, and the photos start to circulate, the complex temporality of the destruction is revealed. Sometimes, there is a "more" to nature than our eyes can see.

Scene 2: Crossing Worlds with Ghost Gum Trees

Moving to another South, in Australia trees are also incorporated as witnesses of ecological urgency. In January 2013, two ghost gum trees, in the Northern Australian district of Alice Springs, were set on fire.[11] For the local Indigenous population, these trees are sacred and seen as living spirits, and, as with the Rio Doce in Brazil, the Indigenous populations of Alice Springs are kin with them. The trees are also ancestors, not just connecting people to the land, but serving as their elders and telling their histories. However, these were not ordinary trees, as they were in the process of becoming a cultural national heritage and were, to use anthropologist Marisol de la Cadena's term, "moving across" (116) an Indigenous world toward a non-Indigenous world. In her research among Andean shamans in the Cuzco region, De la Cadena describes how certain shamans act as translators and mediators between an Indigenous perspective and a modern political tradition. De la Cadena identifies this mediation process as a political strategy of survival and sees it as a real movement across different worlds: the Indigenous and the modern. In De la Cadena's terms, "earth-beings," such as mountain spirits, are called to intervene in and influence a political scene. In this sense, the Andean shamans must manage these forces and their interaction within a modern political and economic arena, acting as translators between what seems to be two irreconcilable worlds. In the case of Alice Springs, the ghost gum trees were in the process of becoming part of the Australian cultural heritage through a series of complex relations where they were one of the "social actors."[12] The trees, who for the Indigenous populations of Alice Springs share a similar spirituality (and thus humanity) of Andean "earth-beings," would now inhabit two worlds: an Indigenous one, where they act as spiritual ancestry; and a modern one, where they act as Australian cultural heritage.[13]

The reason these specific trees were in the process of becoming a national cultural heritage is that they were featured in Albert Namatjira's (1902–59) famous watercolor series that became an iconic image of the Central Australian landscape, if not of Australia as a whole. Namatjira is considered the first recognized Indigenous artist in the country. He was raised in a Lutheran mission in Central Australia, where he learned the watercolor technique with Australian artist Rex Battarbee.[14] According to Giffard-Foret (31), Namatjira's life trajectory incorporates many traces of a postcolonial cultural war. The *savage* Indigenous artisan who is capable of transitioning to a White society by mastering European artistic technique; the melancholy of an Indigenous artist who in spite of being able to paint like White Europeans is never really accepted or seen as an artist; the tale of civilizing and conversion making Namatjira the good tamed Indian; are a few of the themes that appear in a public discussion over the artist's life. Mainly, a critical cultural theory concentrates on how Namatjira occupies the tragic and complex space between Indigenous and European culture, within the violent process of settler colonialism in Australia and the deterritorialization of Indigenous groups.[15]

Terms such as "mediator," "negotiator," "third space," "wanderer," and "cultural appropriation" can be good analytical operators to think about the meanings of Namatjira's life and work from a sociocultural perspective. However, as Giffard-Foret argues, these terms ignore the aesthetic aspects of the paintings (35). In a closer analysis, Giffard-Foret sees more than a perfect copy of European technique in the paintings and identifies a sensible disorder that destabilizes the Western gaze on nature, much closer to the artistic practices of contemporary Indigenous artists:

> In Namatjira's *Ghost Gum* series, trees do not end up as an ornamental device framing the canvas and directing the viewers' eyes toward a focal point, as is usually the case in Western painting.... It lingers as a shadowy presence, as if detached from the scenery and overbearing at once. Namatjira's mastery resides in the care for details with which he painted it, the folds and knots of its sinewy boughs bearing anthropomorphic qualities. The quasi-human appearance of the tree's texture might be a way for Namatjira of staging himself as embodied in the tree. (37)

The *Ghost Gum* paintings are made using a Western watercolor technique that is constitutive of the materiality of the paintings. However, the colors of the landscape and the uncanny corporality of the trees move away from a simple representation of nature. In Giffard-Foret's analysis, the trees are not framing a landscape that illustrates a human point of view of nature; they are what engage the eyes and center of the paintings. The human-like trees also look back at us, acting as a crossroads between worlds. In this sense, the paintings can be seen as translators promoting a movement "across worlds" (De La Cadena 4).

The paintings decenter a Western gaze in breaking a contemplative relation to nature that is no longer only a background. At the same time, the canvas brings to the fore a corporality that gives the ghost gum trees a body, making it no longer an object of contemplation, but a portrayed subject. Thinking with Kopenawa and the xapiris, the apparition of this body in the paintings reveals ancestry and kin that are invisible to the Western gaze, a virtuality present in the trees themselves and in the landscape as seen by Namatjira. If these paintings are not disruptive or become disruptive through a series of agencies, why is it that, in 2013, more than fifty years after Namatjira's death, people felt the need to burn the real trees down?

From a Western modern perspective, the burned-down trees can be seen as a crime against the recognition of Indigenous cultural production or a continued colonial violence against Indigenous sacred territories. However, there is also the disruptive nature of the paintings, which, more than illustrate, give a human-like body to these sacred trees. At the same time, the *Ghost Gum* series desacralizes colonial divisions by connecting the worlds—Indigenous and modern—colonial power has worked so hard to keep separate. What the burning of the trees reveals is how a colonial perspective is still active. The cultural heritage process confirmed the importance of the gum trees, creating a continuity between the trees on the canvas and the trees on the ground, a continuity between Natmajira's Indigenous perspective and the materiality of the real trees. This continuity, from a colonial perspective, was unbearable, and the real trees needed to be burned down.

Scene 3: Tree Ghosts Are All That Remain

The final scene is also set in Australia. It is the summer of 2019/20, and Australia caught fire at an intensity never seen before. According to the "Monash Climate Change Communication Research Hub" report on the fires, "by the end of January 11 million hectares of bush, forests and parks had burnt nationally" (Burgess et al. 5). The dimensions of the bush fires, in terms of space and time, and the destruction were impressive. Many scientists warned that the big fire was coming, but, even for them, it felt as if it came sooner than expected. One dramatic weather event that occurred in the Australian bush fires was "fire-induced and smoke-infused thunderstorms" that scientists call *pyrocumulonimbus* (Peterson et al. 1). These storms, induced by the particles of smoke in the air, produced thunder and sparked new fires. The fires were, to some extent, feeding themselves new fires through these storms. The quantity of these storms in Australia, during the passage from 2019 to 2020, was so significant that scientists identified their traces in the stratosphere as "roughly 1.0 Tg of cumulative smoke particle mass being injected into the lower stratosphere, consistent in magnitude with the initial ash and sulphate plume of a moderate volcanic eruption" (1).

Media coverage of the Australian fires also evoked an image of the apocalypse: "Apocalypse Comes to Kangaroo Island," "The Banality of Apocalypse: Escaping the Australian Fire," and "Apocalyptical Photos from the Frontline of Australia's Bushfires" were some of the headlines at the time.[16] The multiple days of fires came to be known in media coverage as the "Black Summer," and a study shows how the second most common narrative theme in the main media channels regarding the fires was "unstoppable power of nature" (Burgess et al. 23):

> This narrative focuses on how powerless humans are in the face of disaster events. The lack of control we have over disasters is generally juxtaposed by the destructive power of the extreme weather events.

> In the Black Summer, the destruction of fires was highlighted with first person accounts of the fires and descriptions of the destruction caused. (23)

The same study also claims that the main narrative theme in media coverage of the Black Summer was "triumph of humanity" (12), which connects to a common media strategy regarding climate change focused on the human aspect of catastrophes.[17] However, this is not always the case. The headlines of a story published on January 15, 2020, by *Vice* says: "'Tree Ghosts' Are All That Remain in Parts of Burnt-out Australia."[18] In the story, the journalist Gavin Butler narrates his journey through the aftermath of fires in New South Wales, ten days after the events of the Black Summer. The story begins, as many related to the fires, describing the desolate destruction with charcoaled skeletons of burned trees scattered throughout the landscape. This is until Butler begins to notice a different mark in the bush, as he recalls, "not quite trees but the outlines of trees, stark white silhouettes printed like x-rays onto the bone-dry soil."[19]

Not knowing what he was looking at, Butler got an answer from his guide, who said, "Tree ghosts," and added, "Aren't they magnificent?" After consulting specialists, Butler informs us in his article that these marks were caused by the extreme temperatures and duration of fires as well as the dry conditions of the land that, in the exceptional context of the Australian fires, printed these tree negatives on the ground. It is as if white men had to burn down trees to their core to see the souls that Indigenous populations have talked about for centuries. The ghosts that were somehow present in Namatjira's paintings and haunted white men to the point they had to burn down the real gum trees were exposed on the ground and can now be seen in these photographs.

The photographs of tree ghosts mediate the insurgence of the trees' missing body that used to encapsulate the now exposed ghost—the now exposed violence of the Anthropocene, again joining the past of the tree's spiritual ancestry to a future apocalypse. The fire acts upon the trees, burning them at extremely high temperatures so only these lines survive on the ground. However, the trees are also an agent, by marking on the ground their existence that later will be photographed. These trees, destined to disappear in the all-consuming fire, persist in the marks and on the photographs as a ghostly presence. Contrary to Namatjira's ghost gum trees, these trees left a mark that evokes not only the temporalities of the Anthropocene's haunted landscape, but also the materiality of a deadly future. Could we look at these tree ghosts as the revenge of Namatjira's burned-down ghost gum trees? Can these ghosts be the memory and bear witness of the fire that consumed everything?

Conclusions: Aren't They Magnificent!

Against a linear concept of history that reinforces power structures and ignores social struggle, Walter Benjamin says that the past "can be seized only as an image that flashes up at the moment of its recognizability and is never seen again" (390). This means that something in the now (e.g., a body, an image) erupts, revealing uneasiness and singularities (i.e., discomfort)—symptoms usually kept invisible. This image or body does not appear from anything or from an empty space but rather is a virtuality, deeply buried in the present, that needs to be brushed against the grain to become visible. Could the tree ghosts of Australia's New South Wales and the thick red line in the Brazilian photographs be as Benjamin's flash of light—the critical point of history or, to use a more ecological term, the "tipping point" of history?

Looking at these images from Brazil and Australia, the physicality of what survives and produces images before there is even a photograph—the tree bodies and their souls—points to a territory that is crossed by a body that has suffered the long colonial violence and survived in spite of all. The images reveal a nature composed of many bodies and different ontologies—of the Krenak, Yanomami, and Indigenous Australians that make the spirit and body of these trees visible and legible. In his analysis of Benjamin's Angelus Novus allegory, Antonio Negri says that it is "not a theology of the past, but an ontology of the present, of the not-yet" (42).[20] The images analyzed here, which claim a body for these trees, are also claiming the presence of neglected ontologies, massacred by the Plantationocene—the xapiris who are made visible by the tree's printed souls on the devasted land.

There is also a survival that can be seen through these trees, marked by toxic mud or unruly fires. Even the ghost, which appears marked on the ground of the aftermath of the Australian fires when there seems to be nothing left, is a presence and a reminder that there is more than "nature" in these damaged landscapes. This relates to Anna Lowenhaupt Tsing's seminal question based on her ethnography on mushroom pickers: How do we survive in a ruined landscape? (18). Tsing looks at an economy built around a specific mushroom that grows in devasted landscapes and is very appreciated in certain parts of the world. To be able to profit from this improbable life form, which grows where there should be no more life, you must first learn how to notice life amid the ruins. To think about the art of mushroom picking is a way of learning how to survive: "We are stuck with the problem of living despite economic and ecological ruination. Neither tales of progress nor ruin tell us how to think about collaborative survival. It is time to pay attention to mushroom picking. Not that this will save us—but it might open our imagination" (19). The bodies and ghosts of these trees, magnificent as they can be, are a possibility of survival in an endangered world—or, at least, to see them, beyond the destruction they suffered, is to open our imagination to other worlds and ontologies.

Acknowledgments

This chapter received funding from the CAPES/DAAD PROBRAL cooperation project "Discomforting Territories: Images, Narratives, and Objects of the Global South," hosted by Tübingen University and Universidade Federal Fluminense.

Notes

1. On the science wars regarding Global Warming see chapter 6 of *Merchants of Doubts* (Oreskes and Conway), where the authors describe how scientists backed by big oil corporations worked to question the data from climate scientists raising doubts on who and what was responsible for global warming.
2. For a more thorough discussion on the controversies of the term Anthropocene, see Bruno Latour's fourth lecture *The Anthropocene and the Destruction (of the Image) of the Globe* in *Facing Gaia: Eight Lectures on the New Climate Regime* (111–45).
3. The Yanomami are an Indigenous group that occupies a territory in the borders between Brazil and Venezuela in the Amazon Forest. The Yanomami lived almost isolated until the mid-1970s, when the building of highways brought an intense population of gold prospectors into their territories. The Yanomami Indigenous territory was only recognized by the Brazilian government in the 1990s, holding back temporarily the invasion of the territory by gold prospectors. However, the threat of a new wave of gold mining has been intensified in the last years.
4. Kopenawa contests our definition of nature and opens our eyes to other understandings of what "nature" can be. The discussions on a definition of the concept of nature is very well developed by Latour in his lecture "On the Instability of the (Notion of) Nature" (8–40). Also see Edurado Viveiros de Castro's discussion on the Amerindian "multinaturalist" perspective, of a plurality of natures, present in his

seminal essay "Perspectival Anthropology and the Method of Controlled Equivocation." In this debate, there are many forms of defining, experimenting, and thinking about nature. When we say "nature" in this chapter, we are thinking of the modern conception of a natural world that is separated from humankind, a separation we argue certain images contest. When we stress "something more" than nature, or "nature as White Man calls it," we are highlighting the many possibilities of the concept nature.

5   Viveiros de Castro, in the preface of the Portuguese version of *The falling sky*, affirms that the book is addressed to the majority of the Brazilian population called "white" by Kopenawa (2019, 8). Viveiros also clarifies how the Yanomami term *napë*, "originally used to define a relational and mutable condition of the 'enemy, began to have as a prototypical referent the 'White,' that is, members of (any colour) those national societies that destroyed political autonomy and economical sufficiency of the native reference people" (25) (freely translated from: "originalmente utilizado para definir a condição relacional e mutável de 'inimigo,' passou a ter como referente prototípico os 'Brancos,' isto é, os membros [de qualquer cor] daquelas sociedades nacionais que destruíram a autonomia política e a suficiência econômica do povo native de referência" [25]).

6   To better understand this virtual multiplicity of Others, Viveiros de Castro suggests we look at Indigenous mythologies, when a precosmological state was composed of a *background of virtual sociality* where each person contained the infinite possibilities that later, in a process of transformation and metamorphosis, will actualize in the many multiple beings (bodies) that compose our present world. This means that there is always a degree of indiscernibility in the relations between ourselves and this multiplicity of Others: "It is strictly impossible to know whether the mythic jaguar, say, is a bundle of human affects in jaguar shape or a bundle of feline affects in human shape, since mythic metamorphosis is an event or a becoming (an intensive superposition of states) not a 'process' of 'change' (an extensive transposition of states). . . . Absolute transparency bifurcates at this point into a relative *invisibility* (the soul) and *opacity* (the body)–relative because reversible, since the virtual background is indestructible or inexhaustible" (*The Relative*, 9).

7   To "make kin" is a provocation by Haraway of a possible assemblage between all "critters" (including humans) in the world as a form of addressing the urgency of our ecological crisis: "All critters share a common 'flesh,' laterally, semiotically, and genealogically. Ancestors turn out to be very interesting strangers; kin are unfamiliar (outside what we thought was family or gens), uncanny, haunting, active" (103).

8   See the image of the site of Deserto Vermelho on TOBE Gallery's website: https://www.tobegallery.hu/tobe-gallery-artists/bruno-veiga.

9   The essay was published online at Revista ZUM and can be accessed at https://revistazum.com.br/ensaios/mariana-mg-bruno-veiga/.

10 Henri Cartier-Bresson, one of the most important iconic photographers of the twentieth century, helped to shape an idea of photographicspread"ctice. He defined the art of photography by the ability of the photographer to capture a "'fraction of

a second": "To me, photography is the simultaneous recognition, in a fraction of a second, of the significance of an event as well as of a precise organization of forms which give that event its proper expression" (quoted in Hermason 31).
11 Ghost gum is a typical tree of Central Australia; the name "ghost" comes from the effect of the smooth white bark that reflects the moonlight.
12 Here I follow Gell's theory of agency, where "things" can be seen as social agents depending on the context. "Social agency can be exercised relative to 'things' and social agency can be exercised by 'things' and also animals" (17).
13 Marisol de la Cadena highlights that "practices with earth-beings do not necessarily follow distinctions between the physical and the metaphysical, the spiritual and the material, nature and human" (25).
14 For more detailed information on Namatjira's life and legacy, see Martin Edmond's "Battarbee and Namatjira," which concentrates on the ambiguities and complexities of the relation between the colonizer tutor and the subaltern student who excels the master.
15 See Albert Namatjira's *Ghost Gum* tree painting and the comparison with the burned trees in *Australian Geographic*: https://www.australiangeographic.com.au/news/2013/01/albert-namatjiras-ghost-gums-burned-down/.
16 Seen on October 2021 at https://www.bbc.com/news/world-australia-51102658#; https://theintercept.com/2020/01/01/banality-apocalypse-australian-fire/; and https://www.vice.com/en/article/akwdp5/photos-frontline-of-australian-bushfires.
17 For example, the British newspaper *The Guardian* has published guidelines for photographers covering catastrophes and the effects of climate change, indicating where they should avoid abstract photos of nature and concentrate on human suffering. The change is based on NGO "Climate Visuals" research that argues, after empiric research, that people are saturated with images of the consequences of climate change to wildlife. As the newspaper argues in relation to images regarding global warming, "an image of a polar bear on melting ice has been the obvious–though not necessarily appropriate—choice." The problem with the polar bear photo, the newspaper concludes, is that it is a story, but not a "human one." Accessed in October 2021 at https://www.theguardian.com/environment/2019/oct/18/guardian-climate-pledge-2019-images-pictures-guidelines.
18 Story can be accessed on Vice at https://www.vice.com/en/article/pkepdn/tree-ghosts-remain-burnt-out-australia-bushfires.
19 Photo of tree ghost in South New Wales, *Vice,* October 2021, https://www.vice.com/en/article/pkepdn/tree-ghosts-remain-burnt-out-australia-bushfires.
20 The Angelus Novus is an image inspired by a Paul Klee painting used by Benjamin as an allegory for history. In the painting, there is "an angel who seems to move away from something he stares at. His eyes are wide, his mouth open, his wings spread" (392). He looks back at the past, a ruin of cumulative catastrophes, but is blown toward the future by the "storm of progress" (392).

Works Cited

Benjamin, Walter. "On the Concept of History." *Walter Benjamin Selected Writings: Volume 4, 1938–1940,* edited by Howard Eiland and Michael W. Jennings. Belknap Press of Harvard UP, 2006.

Burgess, Tahnee, James R. Burgmann, Stephanie Hall, David Holmes, and Elizabeth Turner. *Black Summer: Australian Newspaper Reporting on the Nation's Worst Bushfire Season, Monash Climate Change Communication Research Hub.* Monash UP, 2020.

Chakrabarty, Dipesh. "The Climate of History: Four Theses." *Critical Inquiry,* vol. 35, no. 2, Winter 2009, pp. 197–222.

De la Cadena, Marisol. *Earth Beings: Ecologies of Practice Across Andean Worlds.* Duke UP, 2015.

Edmond, Martin. *Battarbee and Namatjira.* Giramondo, 2014.

Gell, Alfred. *Art and Agency: An Anthropological Theory.* Clarendon, 2013.

Giffard-Foret, Paul. "Settling Scores: Albert Namatjira's Legacy." *Commonwealth Essays and Studies,* vol. 41, no. 1, 2018, pp. 31–42.

Gómez-Barris, Macarena. *The Extractive Zone: Social Ecologies and Decolonial Perspectives.* Duke UP, 2017.

Haraway, Donna J. *Staying with the Trouble: Making Kin in the Chthulucene.* Duke UP, 2016.

Haraway, Donna J., Noburo Ishikawa, Scott F. Gilbert, Kenneth Olwig, Anna Tsing, and Nils Budandt. "Anthropologists Are Talking—About the Anthropocene." *Ethos,* vol. 81, no. 4, 2016, pp. 535–64.

Hermanson, Sarah. "Henri Cartier-Bresson." *MoMa,* vol. 3, no. 6, 2000, pp. 31–32.

Kopenawa, Davi, and Bruce Albert. *The Falling Sky: Words of a Yanomami Shaman.* Belknap P of Harvard UP, 2013.

Kopenawa, Davi, and Bruce Albert. *A Queda do Céu: Palavras de um Xamã Yanomami.* Companhia das Letras, 2019.

Latour, Bruno. *Facing Gaia: Eight Lectures on the New Climatic Regime.* Polity, 2017.

Mirzoeff, Nicholas. *The Right to Look: A Counterhistory of Visuality.* Duke UP, 2011.

Mondzain, Marie-José. "Can Images Kill?" *Critical Inquiry,* vol. 36, no. 1, 2009, pp. 20–51.

Negri, Antonio. "Uprising As an Event." *Uprisings,* edited by Georges Didi-Huberman. Gallimard/Jeu de Paume, 2016, pp. 37–45.

Nixon, Rob. *Slow Violence and the Environmentalism of the Poor.* Harvard UP, 2011.

Oreskes, Naomi, and Erik M. Conway. *Merchants of Doubt: How a Handful of Scientists Obscured the Truth from Tobacco Smoke to Global Warming.* Bloomsbury, 2010.

Peterson, David A., et al. "Australia's Black Summer Pyrocumulonimbus Super Outbreak Reveals Potential for Increasingly Extreme Stratospheric Smoke Events." *NPJ Climate Atmosphere Science,* vol. 4, no. 38, 2021, doi:10.1038/s41612-021-00192-9.

Silva, Camila, Ane Alencar, Aline Pontes, Julia Shimbo, and Wallace Silva. *The Hidden Emissions: How Amazon Wildfires Can Boost Brazil's CO2 Emissions*. Policy brief, Instituto de Pesquisa Ambiental da Amazônia, Brasilia, November 5, 2021, https://ipam.org.br/wp-content/uploads/2021/11/Policybrief-IPAM-fire.pdf.

Tsing, Anna Lowenhaupt. *The Mushroom at the End of the World: On the Possibility of Life in Capitalist Ruins*. Princeton UP, 2015.

Veiga, Bruno, and Eduardo Sterzi. "Fotografia e Catástrofe: Mariana (MG)." *Revista ZUM*, June 3, 2016, https://revistazum.com.br/ensaios/mariana-mg-bruno-veiga/.

Viveiros de Castro, Eduardo. "Perspectival Anthropology and the Method of Controlled Equivocation." *Tipití: Journal of the Society of the Anthropology of Lowland South America*, vol. 2, no. 1, 2004, pp. 3–20.

Viveiros de Castro, Eduardo. Preface, "O Recado da Mata," to *A queda do céu: palavras de um xamã Yanomami*, by David Kopenawa and Bruce Albert, Companhia das Letras, 2019, pp. 7–28.

Viveiros de Castro, Eduardo. *The Relative Native: Essays on Indigenous Conceptual Worlds*. Hau, 2015.

9

Toxic Transits

Ghostly Double Gazes, Slow Violence, and North-South Ecologies of Inequalities in the Films *Sealed Cargo* (Bolivia) and *Arica* (Chile)

Azucena Castro

Out of Sight, Out of Mind: Ghosts in Somebody Else's Backyard

The dumping of toxic waste in the Global South has largely remained a hidden phenomenon from public and media attention. Still, since 1990, it has received increasing cultural, environmental, social, and legal focus.[1] Transborder toxic disposal in the Global South intensified in the 1980s with the enforcement of environmental regulations in the Global North, which forced companies to find disposal sites "out of sight—and out of mind" (Melosi 17). This neocolonial praxis was further legitimized by global institutions such as the World Bank, whose leaders declared that underdeveloped countries did not produce enough waste, and, therefore, it was necessary to redistribute toxic waste from the North to the South (Nixon 1). Toxicants leave such diffuse traces that environmental humanities scholar Rob Nixon speaks of "slow-motion toxicity" (3) to suggest the pervasive but deferred violence that toxicity produces on matter and bodies.

Tons of toxic minerals, hazardous rubbish, and e-waste from industrial processes have been shipped to China, Ghana, India, and Latin America from the Global North. According to Ghanaian environmental journalist and activist Mike Anane, this practice constitutes a crime against humanity and the environment (*The E-Waste Tragedy*).[2] This chapter focuses on the cinematic rendering of toxic waste dumping in Global South geographies and atmospheres. The environmental movies *Sealed Cargo* (*Carga sellada*, 2015), a fiction film by Bolivian director Julia Vargas-Weise, and *Arica* (2020), a documentary by Swedish film directors Lars Edman, a Chilean Swede from the Chilean diaspora in Sweden; and William Johansson Kalén, offer unique cinematic explorations of the undisclosed practice of global waste dumping in Latin America during the 1980s and 1990s. While studies of relations between cinema and the environment in Latin America have mainly focused on how Indigenous cinema frames popular environmentalisms (Marcone), how rural landscapes are portrayed as exhausted places in the background of ecological devastation (Andermann), and, more recently, how ecocinema decenters the human facing climate change events (Fornoff and Heffes), my chapter will contribute to this scholarship with insights into the underexplored theme of how cinema portrays global toxic waste dumping in a North-South perspective. I argue that these two movies depict and critique the hidden and deferred effects of global toxic waste dumped in the Bolivian Highlands and the desert town of Arica in the Atacama region of Chile. What is the role of environmental filmmaking in the context of the global waste economy? How can contemporary cinema create stories relevant to the communities and bodies affected by toxicity while connecting to social, economic, legal, and environmental questions?

Environmental Racism, Ghost Acres, and Deadly Affairs in Cinema

The transboundary shipment of hazardous waste from the Global North to the Global South is intersected by the question of "environmental racism," a term first coined by Bullard referring to the "deliberate targeting of communities of color for toxic waste facilities" (Okafor-Yarwood and Adewumi 286). The tougher environmental regulations in Global North countries (as I will discuss with respect to *Arica*) and the reception of international waste by Global South governments (as I will elaborate on for *Sealed Cargo*) have settled this colonial and racist practice. Toxic waste dumping has two correlations. First, toxicity and pollution are associated with the long wake of colonialism (Liboiron "Waste Colonialisms"), including the persistence of racialized hierarchies that make it "acceptable" for powerful, mostly white, and prosperous decision-makers to dump in Black communities, for example, both in the United States (continuously marginalized and excluded from participatory decision-making, despite living in a democratic country in the Global North) and in the poorer areas of the Global South. Second, neoliberal globalization and fossil capitalism made the transatlantic transport facility possible. To comply with tougher environmental laws that protect the environment at home, Global North countries have sought places to dump industrial disposals in the Global South, taking advantage of corrupt or negligent governments that have accepted the toxic cargo in exchange for payment. Despite the legal and illegal ways to ship toxic loads to the Global South, these are still, as Antonia Alampi contends, "deadly affairs" (11) that produce contamination and disease in remote destinations. However, I propose that the treatment of global toxic waste in Vargas-Weise and Edman and Kalén's movies question Alampi's suggestion that it is "impossible to contain its advancement" (11), and, therefore, that the only thing left for us (and art) to do is to "embrace, understand and engage" (11). Instead, as I will develop here, these cinematic practices also offer activist responses, thus calling for a toxic justice to stop the modern machinery of progress that produces toxicity.

The manifold techniques of legal circumvention employed by corporations in the Global North to get rid of toxic waste in the Global South have brought to light understandings of the spaces chosen for hazardous waste as "ghost acres" (Müller, "Hazardous Travels" 29):[3] a displaced consumption of resources and hazardous waste dumping in remote places used as wasteland within structures of "garbage imperialism."[4] The global production and transboundary transport of hazardous material expose North-South inequalities prevailing in global waste management and encourage rethinking the planet's natural and cultural resources as "toxic commons" when all habitats become increasingly hazardous (Müller, "Toxic Commons"). Focusing on toxic bodies as porous spaces, Stacy Alaimo (22) claims that bodies are "trans-corporeal spaces" traversed by all types of substances—for example, toxic water, air, and soil, thus defying the idea of bounded individuals and contained spaces (22). Focusing on recent Latin American literary productions dealing with agrochemical toxicity in rural areas, Gisela Heffes ("Escrituras tóxicas" 365) identifies a "toxic continuum" that connects urban and rural zones through metamorphosed, spectral, and ghostly bodies that represent the crossing of the last toxic boundary. These toxic boundaries form spectral spaces that Elaine Gan and Anna Tsing et al. call ghosts of haunted landscapes of the Anthropocene: "the traces of more-than-human histories through which ecologies are made and unmade" (G1).

The Uyuni area in the Bolivian Highlands and Arica in the Atacama Desert, as registered in *Arica* and *Sealed Cargo,* present two crucial sites for understanding slow violence and environmental racism in the context of global hazardous waste dumping. The movies expose the violence against the recipients of global toxicity. This chapter offers a cultural and environmental analysis of Vargas-Weise's and Edman and Kalén's movies, which elaborate on the logic of global waste disposal in Latin American geographies and atmospheres.[5] I approach toxicity in the movies as a trace of hazardous materials that connect matter, bodies, and environments (Alaimo), but also I explore how the movies portray the realization that the dump also pervades moviemaking as an "ongoing and morphing process" of toxic film semantics. This double perspective affords a way to understand how toxicity blurs internal and external boundaries, thus constituting a social, health, and cultural issue. Jorge Marcone (209) contends that environmental documentaries often thrive on portraying social and environmental conflicts affecting the daily lives of human communities and the local resistance, but these documentaries fail to convey more-than-human ontologies and Indigenous temporalities. As I will develop, Vargas-Weise's and Edman and Kalén's movies deploy experimental techniques to signal the toxicity in matter and bodies that remains out of sight while also foregrounding local communities and Indigenous epistemologies.

In the 1990s, toxicity entered global environmental and health discourses when the World Health Organization (1993) warned that toxicants were widely spread even in privileged regions, where the most toxic residues were produced. However, questions of power and inequality still remain regarding toxic waste, since Global South countries receive a significant influx of toxic material every year from the industrialized Global North. As environmental history scholarship has proposed, contaminated communities have also been "exposed to narrative violence consisting in the silencing and invisibilization of their stories" (Armiero et al. 8). The narratives that seek to render visible the effects of toxicity also expose the biographies of contamination and the resistance of affected communities. As cinematic narratives on toxicity, I argue that *Sealed Cargo* and *Arica* configure *ghostly double gazes* that explore toxicity traces in the materiality of bodies and landscapes by negotiating between "transit" and "emplacement" through a particular cinematic rhetoric that connects visuality to embodiment and ghostliness. In this sense, these movies offer ways to observe not only how cinema responds to the traffic of hazardous materials by foregrounding affected bodies and their narratives, but also how the cinematic medium uncovers the violence of colonial discourses by emphasizing the ghostliness of toxicity: seemingly invisible toxic effects that emerge over long-term exposure. To examine how these ghostly double gazes are configured in the selected movies, I combine environmental humanities, environmental justice, and environmental history. These perspectives provide tools to explore how film as a cultural product generates embodied approaches to toxic bodies; how the movies engage the right to a healthy environment, acknowledging the disruption of cultural and ancestral relations caused by environmental racism; and how ghostly relations are exposed as hidden narratives of poisoning that disrupt the lineal temporality of history in the movies.

Arica: Summoning the Ghost of Toxic Pasts

In 1984, Swedish mining company Boliden in Rönnskär shipped a toxic cargo of over twenty thousand tons of waste containing arsenic and lead, among other metals, to be dumped in Arica, Chile, through an international contract with the Chilean company Promel, which was to receive and neutralize the toxic metals.[6] However, the piles of arsenic and lead lay bare and untreated until 1998, while population growth led to the construction of a state-sponsored neighborhood called Polígono beside the untreated hazardous waste. Children played in the piles of arsenic and lead, and families inhaled the air charged with polymetallic particles in this highly windy area. The demands from affected families prompted a lawsuit against the Chilean State and Promel due to the deaths, genetic malformations, and stomach and lung cancer cases connected with the chemical dump. While Arica and the "polymetallic zone," as it is known in Chile, was a matter of discussion in the Chilean public arena, in Sweden the event had largely vanished from public memory until 2006, when Swedish filmmakers Lars Edman and William Johansson Kalén visited Arica on a trip related to their film studies.[7] On this trip, they became acquainted with the situation and decided to investigate it through film.

The visit to Arica prompted the shooting of the first documentary *Blybarnen* (*Toxic Playground,* 2010), which draws attention to the disastrous impact of the toxic cargo shipped from their homeland, Sweden, to Chile, twenty-six years ago. Edman was born in Chile, but, as a child, he was adopted by a Swedish family and grew up in the Swedish village of Skellefteå, also Kalén's birthplace, where the mining company Boliden was established in 1924. The first documentary brought public and media attention to this deadly affair that had remained undisclosed for Swedish citizens. It prompted research by environmental lawyers and an international lawsuit against Boliden, the

Figure 9.1. Aerial view of Arica town in Chile. Film still. *Arica*.

first of its kind in Sweden and Scandinavia. In 2020, Edman and Kalén released the sequel documentary *Arica,* focusing on the legal processes of that lawsuit, where environmental lawyers from Sweden and the United States were also involved.

The movie's opening scene shows the courtroom where the hearing on Arica's case occurred. We hear off-screen recordings of the prosecutor while we see a panoramic image of Rönnskärsverken on the outskirts of Skellefteå, followed by the image of a cargo train bearing the inscription "Boliden" placed at the entrances of the company's headquarters, suggesting the transit of materials to an outside location. This chain of images (courtroom, the industrial plant, and transportation) poses questions about where these materials are being transported to and what environmental, health, and legal implications this transport has, thus opening a space to render visible the shipped toxicity. The camera constantly shifts from close-ups focusing on Arica and affected inhabitants to landscape shots over Arica and Skellefteå. These shifts connecting emplacement to transit entangle the human toxic stories with the unequal environmental histories of these two remote places. The combination of panoramic shots over Arica and Boliden's headquarters also intends to engage a Swedish audience for whom the toxic transit of the waste produced in Sweden passed unnoticed.

Figures 9.1 and 9.2 are repeated throughout the movie with angle variations, thus producing a dialogue between situatedness and transit. This cin-

Figure 9.2. Aerial view of Boliden's headquarters in Sweden. Film still. *Arica.*

ematic practice links Sweden-Chile and Boliden-Arica through an invisible toxic link tying North and South. This form of image-making produces a ghostly link that puts into vision the practice of toxic colonialism in the involvement of Global North companies in toxic waste dumping in the Global South. Arica's inhabitants and their toxic bios emerge in the movie as the reverse image of the modern imaginary in Sweden, "ghost acres" (Müller 2019), an odd, extraterritorial extension or ghostly backside of the so-called developed world. As a symbolic place, *Arica* unravels ghostly presences (sick, contaminated) in this marginal overseas place that remained in the shadows while receiving the violence of the industrialized machinery. If, as Gan et al. (G2) suggest, "every landscape is haunted by past ways of life," *Arica* summons the lives and deaths in these ghost acres through shots of the local graveyard, diseased people, and contaminated soil that haunt Chile's and Sweden's modernity.

Furthermore, *Arica*'s aerial shots are combined with experimental footage in the Polígono and Skellefteå neighborhoods, calling attention to inequalities regarding access to a healthy environment. For example, the aerial shots of a pine tree forest, close-ups of a bird in the snow, panoramic views of residential districts in winter and the bay where Boliden has its headquarters in Sweden contrast with aerial images of the cemetery in Arica and footage taken with a handheld camera from a moving vehicle in Polígono showing precarious residences. Through this strategy, the film produces a perceptive and sensorial proximity that challenges the geographical distance between these two places. In fact, the film's narrator, Lars Edman, expresses the following: "The distance is big between Arica and Boliden. Reality on the other side of the Earth is far away. Almost absolutely quiet" (my translation). This image-making situates Arica and Skellefteå horizontally, thus challenging the erasure of where the waste is deposited—dynamics that characterize global waste management—to reveal the trail of toxicity that entangles the environmental histories of these seemingly faraway places.

The juxtaposition of archival images and audio from the 1980s with present-day footage enables the viewer to perceive both the elusive traces of toxicity and the unequal North-South relations. The scene in which the environmental lawyers visit the place where the toxic waste lay bare until 1998 is interspersed with archival images from Arica in the 1980s. The film recovers archival films, photographs, audio, maps from Arica and Boliden, and images of medical reports on toxicity presented in court by Boliden and Arica inhabitants. These hybrid materials from journalism, law, cartography, and medicine connect the past with the present, prompting the viewer to question the temporalities of toxic matter in the body and justice's shortcomings in considering those temporalities for accountability. In an interview (Castro and Selgas), the filmmakers commented that they intended to show the contrast between Swedish environmental law, where a crime cannot be prosecuted ten years after it was committed if a formal demand was not filed at that time of occurrence, and Chilean law, where it is possible to file a lawsuit for a case when its consequences become visible on bodies or the environment. The interweaving of archival images with present-day shootings offers gazes that uncover the often-obscured human and nonhuman casualties most likely not to be perceived. This twofolded perception in *Arica* helps to visualize the correlations between the affected communities and their habitat as recipients of the toxic material and the remote society in Sweden that produced and shipped the toxicants. This cinematic practice mobilizes the viewer to perceive what was removed from the narrative of modernity, thus reshaping the conditions of visuality in environmental documentaries. *Arica* contaminates the modern narrative of progress in the Global North (Sweden) by exposing the toxic bios of affected humans and degraded places in Arica, hence defying the administration of invisibility by global waste management.

Through a ghostly double gaze, *Arica* places toxicity as both an external and internal phenomenon. The movie deploys experimental strategies by exchanging roles between the observing subject and the observed others and employing anthropology methods of participatory interview and community-based action. In the interviews with Arica inhabitants, Lars Edman exchanges roles with a local girl, Jocelyn, who sometimes holds the microphone, becoming the interviewer and asking Edman why he is in Arica, thus shifting the roles in traditional documentaries. The inclusion of the film director as the interviewed subject blurs the boundaries between the (Swedish) filmmaker and the affected community in Chile; as Edman points out, "I could be on the other side receiving the waste." The interconnection that Edman generates as an outsider-insider informant, or subject-object, attempts to forge transnational solidarity ties that foreground these disenfranchised neighborhoods' plight. These role-shift dynamics suggest that toxicity is not only a matter of toxified bodies but is also profoundly engraved in the lifestyles and worldviews of the Global North. However, as Müller (446) contends, that communal "we" will always remain paradoxical, since, while it evokes the tension of "being both complicit in and affected by global environmental degradation," it also marks the boundaries that determine the degree to which one is exposed and how toxicity is experienced in diverse bodies and geographies.

While the commons are increasingly becoming toxic, the experiences of contamination, material exposure, and risks for disease remain unevenly shared. In her study on disability and the environment, Fritsch (363) analyses the existence of "subclinical toxicity" that is not usually traceable by standard medical procedures but points to a "reduction in intelligence levels and changes in their behavior," especially in children exposed to lead. This context in the United States can be extrapolated to the Arica case. Jocelyn, the interviewed girl from Arica, talks about body malformations, diminished bodily and cognitive functions, and fears about unperceivable risks and dangers of disability in infants in her community. The double cinematic gaze the movie mobilizes points to the visible material traces of toxicity, as well as the invisible and hidden ones, thus yielding what Lawrence Buell (35) calls "images of community disruption" that create a looming sense of threat and risk haunting the joyful memories of well-being, prosperity, and health in the Global North, thus renegotiating the ideas of emplacement and transit, far and near.

As discussed in this section, *Arica* portrays the unequal relation between the sender and the receiver of toxic waste, thus problematizing global toxic waste as a concern only for the affected local community. Instead, the movie sets the spotlight on national governments and the societies that produce and ship away toxicity. Through experimental documentary strategies, *Arica* summons the ghosts of deadly affairs past, thus turning the documentary into an art of tracing the remote bodies and places lying in the shadow of the bright tales of modern progress in the Global North.

Sealed Cargo: Delving into Toxic Rumors of a Ghost Train

Sealed Cargo (2015) is a road movie about a secret train crossing Potosí and Oruro.[8] Cochabamba film director Julia Vargas-Weise developed the idea for the film after a rumor spread among local communities in Oruro in 1994 about a train crossing the Bolivian Highlands to dump a secret cargo of apparently toxic minerals on the frontier with Chile. Although the event was documented by local radio stations at that time ("Concentrados mineralógicos," *Agencia de Noticias Fides*), it remained a rumor, and the event was not further investigated. This train, seen wandering about in the Altiplano, received the popular name of "ghost train" due to the secrecy surrounding the cargo and its destiny. *Sealed Cargo* elaborates on this rumor.

The film's protagonist is a high-ranking police officer, Marshal Hector Mendieta, who lives in La Paz. In exchange for a promotion to be relocated to the United States as a federal agent, he accepts the mission of conducting a train loaded with a secret cargo to be dumped at the Chilean border. On this trip, the Marshal and the three accompanying police officers meet resistance from local communities trying to force the train to return the cargo to its origin. The Marshal is commanded to employ an old local train from Uyuni for the task. The train trip involves several characters that take part in the plot: a local train conductor who calls the train "La Federica"; a country girl called Tania who boards the train trying to escape a forced marriage; and the local communities attempting to stop the train supported by an Indigenous guerrilla led by the character Antonio Urdimala. In what follows, I analyze how the movie constructs a ghostly double gaze that captures the barely perceptible unfolding of transformations in the human bodies that come into contact with the toxic cargo and the local resistance to unequal global waste management.

Unlike *Arica,* which focuses on North-South relations highlighting the toxic cargo's origin, *Sealed Cargo* centers on the national relations between the capital city, La Paz, where government officials accepted the cargo, and the rural regions, where the cargo would be dumped. However, the movie acknowledges the global origin of this toxic cargo in the highly evocative

Figure 9.3. Sign showing Africa or South America as destinations of toxic waste. Film still. *Sealed Cargo.*

opening scene, which shows three people wearing hazmat suits who are loading boxes into a transport vehicle. In completing the loading, one of the loaders looks at the destination sign and changes the label of the destination from "Destination South Africa" to "Destination South America."

This label change from South Africa to South America (Fig. 9.3) suggests that hazardous materials are persistently sent to "South" destinations, Africa or America. The label shifting in this opening scene reflects the concept of "environmental racism" (Bullard; Okafor-Yarwood and Adewumi)—that is, the purposeful targeting of territories in the US South, South Africa, and South America where racialized hierarchies and a (settler-)colonial mindset see the territory of a racialized other as a repository for waste. In 1989, the United Nations Environment Programme Convention used the term "toxic waste colonialism" (Liboiron, "Waste Colonialisms," *Pollution*) to articulate the African nations' concerns about the disposal of hazardous waste by high GDP countries into low GDP countries, stating: "High GDP countries like those in Europe and North America were accessing African land for inexpensive disposal of waste. They were using Africa as a sink" (Liboiron, "Waste Colonialisms"). *Sealed Cargo*'s opening scene parallels Latin America and Africa regarding toxic waste colonialism. These dynamics determine which lives are worth preserving as healthy and protected (free of toxicity and in hazmat suits) and which lives will be sacrificed as toxic dumps.

Like *Arica*, *Sealed Cargo* highlights the conflict between expert/scientific knowledge and local/popular knowledge. The Marshal treats the claims of local communities as rumors or folklore while he considers the expert information from the police force and the government officials as scientific and accurate. However, the repeated assertions from the local communities begin to generate doubts in the Marshal—for example, in the scene when a local priest, together with a group of locals, demands that the Marshal do something about the dumped toxic waste because, as he says, "the people are getting sick." When the train crew, including the Marshal, begins feeling sick (coughing, shortness of breath, skin marks), he is forced to accept the connection with the toxic boxes. In this way, the movie exposes the unfolding of toxicity in human bodies becoming sick in parallel with the testimonies of the rural communities, thus defying the scientific and hegemonic knowledge from urban government officials. This embodied double gaze points to the complexities of visualizing toxicity and the almost imperceptible effects of slow violence on human and nonhuman bodies. The unfolding of toxicity in the Marshal's body brings about a realization (a kind of hamartia of the tragic hero) that he is as contaminated as those local communities; hence, his story will also be erased from official records.

Due to his exposure to the toxic soil, the Marshal starts hallucinating. In these scenes, he remembers the first day at what is probably a private school in La Paz, when all the children arrived with their white or mestizo mothers, while he comes with his Aymara mother, who hides at the entrance of the school and prompts him to say that his mother could not go and that she was the maid. This scene connects with the movie's opening scene, in which the Marshal eats in his wealthy house in La Paz, and his Aymara mother serves the food as the maid. Through these memories, the movie connects the colonial history of the exploitation of Indigenous people with the invisibilizing of the claims of rural communities about the slow violence contaminating their bodies and habitats. The toxic boxes in the movie act then as a kind of "tracer" (Gan et al. M8), uncovering social, ethnic, and power relations. The traces of the past do not return so much as in Derridean hauntology—that is, through absent materialities made present through allusion, but through embodied and toxic memories that haunt the landscapes of the Anthropocene, bringing forth histories of the more-than-human (Gan et al.) in the colonial and racist Latin American modernity.[9]

While there are only a few shots of the green toxic material in the movie, the corrosive effects of this cargo on the bodies connect environmental issues with questions of power and ethnicity in which toxic relations such as racism materialize. In this sense, the toxic cargo not only reveals the "environmental racism" (Bullard; Okafor-Yarwood and Adewumi) involved in the dumping of toxicants on areas occupied by peasants and Indigenous communities

Figure 9.4. Train carrying the toxic cargo across the Altiplano, Bolivia. Film still. *Sealed Cargo.*

but also the connections between the racist and the colonial structures still present in Latin America.

The repeated panoramic landscapes (see fig. 9.4) in *Sealed Cargo* highlight the Andean colors. These provincial landscapes, cut across by the train recovered from Uyuni, evoke the history of progress and modernity in Bolivia.

The train is recovered from the Train Graveyard or Train Cemetery in Uyuni, where abandoned and obsolete trains accumulate as industrial waste amid the breathtaking view of the salty lake. The recovery of this ghost train as obsolete technology and industrial waste revived to carry toxic materials binds the history of global waste management with the history of modernity in Bolivia. The driver named the train "La Federica" after an anarchist coworker, thus actualizing the history of the Anarquistas del Norte in Oruro in the 1930s, where workers involved in salt and mining productions revolted against crisis and hunger. In the nineteenth century, British engineers and local laborers built a railway connection between the local mines stretching from La Paz to the seaports in Chile, where the Bolivian Railroad Company transported metals until the 1940s, when mining declined, and the locomotives were abandoned in that Andean landscape. By recovering La Federica as a ghost train, the movie allows the viewer to perceive a parallel between the lack of benefit for the local communities from mining at the beginning of the twentieth century and the contamination that accrues to local communities by global toxic waste received from industrialized countries in the 1980s and 1990s.

This harsh reality of rural life in Bolivia is addressed in the movie through side narratives. Tania, the girl trying to escape an arranged marriage for economic reasons, comments that "the mine swallowed my father," which hints at the rural towns' dispossession and the forced migration. The toxic cargo in this movie becomes a material and symbolic burden affecting the protagonist, the driver, the policemen, the local communities, and the state that wants to get rid of it. The cargo as a ghostly burden is also projected onto the viewer, who is witnessing its impact through the movie, as it orchestrates the narrative while remaining hidden (unseen). In this way, the movie interweaves transnational dynamics with local human and nonhuman communities (Fornoff and Heffes 11) to foster an ecological consciousness and imagination of possible futures for local communities amid toxicity.

By deploying experimental practices, *Sealed Cargo* portrays an Indigenous guerrilla that employs ancient rituals (dance) and community practices to stop the train. This guerrilla intervenes in the plot through oneiric instances that pollute the linear narrative, the straight line of the train, and the forward march of progress. One such dream-like scene consists of typical Andean Potosí musicians and *danzantes* crossing the whiteness of the Uyuni Salt Lake in a train-like line, carrying and playing traditional musical instruments. Another scene depicts surreal images of Indigenous guerrilla fighters wearing masks and ritualistic clothing to defend the humans and the environment (Fig. 9.5) through the local ritual in honor of the god Tiw from Oruro.

The cited image (Fig. 9.5) is connected to the Andean ritual of *la diablada*, typical of Oruro, a dance representing the conflict between good and evil. In the movie, these oneiric shots are disconnected from the main narrative, thus interrupting the march forward of the train and the progress of the narrative plot. These interludes direct the viewers' gaze to submerged cultures and knowledges resisting colonial and racist erasure from historical records through a kind of Indigenous ontohauntology where the spectral does not

Figure 9.5. Indigenous defenders of the territory. Film still. *Sealed Cargo*.

come from the past to reorganize the present, but manifests as materialities, presences, and bodies that can cross realms, thus bringing to the fore the more-than-human histories suppressed by modernity. Using these oneiric images as cinematic infiltrations, the movie highlights Indigenous, popular, local, and unauthorized storytelling (e.g., rumor and ancient rituals) as legitimate sources of knowledge to understand the dispossession caused by global waste dumping entangled with exploitation and extractivism in Latin America.

After leaving the cargo at the military base, suggesting another official would continue searching for a dumping place in this area, the Marshal escapes from the police forces chasing him and gets lost in the Andean landscape. In the movie's last scene, he ventures into the open scenery, removes his clothes, and becomes a formless shadow blending in with the surroundings. In the very last scene, the Marshal's naked silhouette is projected against a stone in an ambiguous shape. This shadowy, ghostly, and contaminated figure merges with the Andean landscape, thus becoming a haunting testimonial presence that discloses the hidden, colonial, and racist dynamics of toxic waste dumping in Latin America. By foregrounding the resistant bios of affected communities, the movie exposes the toxic discourses sustaining the colonial and racist practices of global toxic transits to the Global South's geographies and atmospheres.

Conclusions

By deploying both documentary and experimental practices that connect a transnational North-South gaze with a focus on local communities, the analyzed environmental films, *Arica* and *Sealed Cargo,* portray the racist and colonial dynamics of the global waste economy. This double visualization highlights the toxic bios of affected communities and bodies while connecting to social, ethnic, legal, and environmental history issues, thus renegotiating the terms of emplacement and transit in light of the deadly shipping of hazardous matter to the Global South. The cinematic portrayal of imminent and long-term risk for local communities requires new ecological understandings of the increasingly toxic and deadly commons (Alampi), the uneven distribution of toxicity between North and South, and the transit channels of toxic waste that link particular emplacements in the Global North and Global South.

By striving to present the invisible toxicity traces on matter and bodies, the movies articulate a ghostly perception of the slow-motion toxicity unfolding in bodies and the environment, the local forms of community resistance (Indigenous guerrilla, local gatherings at Arica), and the transit routes of toxic materials from the Global North. These ghostly double gazes are crucial in these environmental movies dealing with global toxic waste. While *Arica* deploys conventional documentary tropes, it also experiments with role shifting (subject-object) and multitemporality by reversing the documentary's focus on pastness to, instead, project the past haunting the present and the future of Chile's and Bolivia's modernities. By fictionalizing a rumor of a ghost train, *Sealed Cargo*'s director showcases grassroots local resistance and highlights an Indigenous guerrilla that challenges the linear progress of the modern narrative. By exploring global waste dumping through ghostly landscapes and bodies, these environmental films document the links between the global logic of producing sites of inequality and toxicity and the emplaced atmospheres and geographies of toxicity in the Global South. Lastly, these two movies engage the history of inequalities in North-South environmental relations while encouraging environmental activisms that forge solidarity alliances with Indigenous communities, local inhabitants, and the situated more-than-human territories.

Notes

1 Robert Bullard's *Dumping in Dixie* (1990) was a pioneering book focusing on environmental justice activism among African Americans and other minoritized communities in the United States who mobilized against toxics. Buell's essay "Toxic Discourse" (1998) laid the groundwork to approach fictive and nonfictive expressions that document toxicity in bodies and the environment produced in the same country. Based on Rachel Carson's report *Silent Spring* (1962), Buell observed that toxic discourse presents, on the one hand, a "pastoral disruption" (647) in the order of representation and, on the other hand, "totalizing images of a world without refuge from toxic penetration" (648).
2 Anane has actively documented the human, atmospheric, and soil contamination due to the e-waste transported to dumps in Ghana.
3 The concept of "ghost acres" refers to virtual landscapes in poor countries that Europe used to continue materializing production and consumption. In toxicity discourse, and as I use it here, it refers to "externalization mechanisms" integral to an economic system that sustains its growth through the appropriation of cheap wasteland—or "ghost acres" (Müller 29) as dumps. Toxic ghost acres show the "paradox of modern environmentalism" (26) in that while industrial nations invest in greener ecologies, they push their hazardous waste to marginal areas, thus creating a "world of environmental inequity" (26).
4 The term "garbage imperialism" was first used in 1986 by William Tuohy in the *L.A. Times* report "116 Nations Adopt Treaty on Toxic Waste," signaling the international commitment and adherence to the Basel Convention to stop the commerce of toxic waste (Müller 29–30). The related term "toxic colonialism" was coined by environmentalist African leaders to denote the struggles of local communities in environmental conflicts involving transnational capital and waste (Okafor-Yarwood and Adewumi 287).
5 Both films have been featured at international environmental film festivals. *Arica* received the Prize in Environmental Anthropology by *CineEco Seia* 2021 from Portugal and the BIFED 2021–Bozcaada International Festival of Ecological Documentary, while *Sealed Cargo* obtained several international prizes that highlight the environmental theme.
6 Already during the 1970s and 1980s, this industrial site received criticism for unhealthy conditions when employees were exposed to toxic materials such as arsenic. In 1980, Ann-Kristin Hedmark and Gunnar Enqvist's socially oriented mu-

sic included the "Song of Rönnskär," which begins with the verse "I work here in Rönnskär—breathing gas and poison every day."

7 In the seminar "Arica: A Toxic Waste Scandal with 'Our Soil' Panel," organized by the North Troy Environmental Justice Film Festival (2021), Chilean anthropologist and activist Rodrigo Pino Vargas commented on the manifestation of polymetallic contamination in Arica since 2005.

8 I thank *Imagina Films* for lending me a copy of the film in 2020.

9 In *Specters of Marx*, Derrida reads ghosts as an absence from the past that supplements the present. Instead, in Anthropocene hauntology following Gan et al., ghosts are presences that inhabit all life-forms, the spectral presence of ancestors in us as a form of a ghostly symbiosis of geological/biological cohabitation.

Works Cited

Agencias de noticias fide. "concentrados minerologicos," 1994, https://www.noticias-fides.com/nacional/sociedad/concentrados-mineralogicos-277935.

Alaimo, Stacy. *Bodily Natures: Science, Environment, and the Material Self.* Indiana UP, 2010.

Alampi, Antonia. "Deadly Affairs." In "Deadly Affairs: An Art Exhibition About Toxicity," special issue, *Springs: The Rachel Carson Center Review*, edited by Antonia Alampi, vol. 5, 2019, pp. 5–21.

Andermann, Jens. "Exhausted Landscapes: Reframing the Rural in Recent Argentine and Brazilian Films." *Cinema Journal*, vol. 53, no. 2, 2014, pp. 50–70.

Arica. Toxic Waste Scandal. Directed by Lars Edman & William Johansson Kalén. Laika Film & Television, 2020, https://aricafilm.com/.

Armiero, Marco, et al. "Toxic Bios: Toxic Autobiographies—A Public Environmental Humanities Project." *Environmental Justice*, vol. 12, no. 1, 2019, pp. 7–11.

Buell, Lawrence. "Toxic Discourse." *Critical Inquiry*, vol. 24, no. 3, 1998, pp. 639–65.

Bullard, Robert D., *Dumping in Dixie: Race, Class, and Environmental Quality*, Westview, 1994.

Carga Sellada. Directed by Vargas-Weise, Julia. Imagina Films, 2015, https://www.youtube.com/watch?v=2yhvkMb74g4.

Carson, Rachel. *Silent Spring.* Mariner, 2002.

Castro, Azucena, and Gianfranco Selgas. "Open Talk—Global Toxic Waste, Slow Violence, and Contaminated Soil: A Conversation on Arica." Latin American Film Series, Romanska och klassiska institutionen, Stockholm University, May 6, 2021, https://www.youtube.com/watch?v=-GV-6a614NY&t=350s.

Derrida, Jacques. *Specters of Marx. The State of the Debt, the Work of Mourning, and the New International.* Routledge, 2006.

The E-Waste Tragedy. Directed by Cosima Dannoritzer, with Mike Anane. La Sept, RTVE, Televisión Española, Televisió de Catalunya, 2014, https://www.imdb.com/title/tt3804476/.

Fornoff, Carolyn, and Gisela Heffes. *Pushing Past the Human in Latin American Cinema.* State U of New York P, 2021.

Fritsch, Kelly. "Toxic Pregnancies: Speculative Futures, Disabling Environments, and Neoliberal Biocapital." *Disability Studies and the Environmental Humanities: Toward an Eco-Crip Theory,* edited by Sarah Jacquette Ray and Jay Sibara, U of Nebraska P, 2017, pp. 359–80.

Gan, Elaine, et al. "Introduction: Haunted landscapes of the Anthropocene." *Arts of Living in a Damaged Planet.* U of Minnesota P, 2017, pp. G1–G14.

Heffes, Gisela. "Escrituras tóxicas: cuerpos y paisajes alterados." *Tekoporá. Latin América Review of Environmental Humanities and Territorial Studies,* vol. 3, no. 1, 2021, pp. 348–67.

Liboiron, Max. "Waste Colonialims." *Discard Studies,* 2018, https://discardstudies.com/2018/11/01/waste-colonialism/.

Liboiron, Max. *Pollution Is Colonialism.* Duke U P, 2021.

Marcone, Jorge. "Filming the Emergence of Popular Environmentalism in Latin America: Postcolonialism and Buen Vivir." *Global Ecologies and the Environmental Humanities: Postcolonial Approaches,* edited By Elizabeth DeLoughrey, Jill Didur, and Anthony Carrigan. Routledge, 2015, pp. 207–25.

Melosi, Martin V. *Garbage in the Cities: Refuse, Reform, and the Environment (History of the Urban Environment).* U of Pittsburgh P, 2004.

Müller, Simone. "Hazardous Travels: Ghost Acres and the Global Waste Economy." *Deadly Affairs,* edited by Antonia Alampi, vol. 5, 2019, pp. 23–32.

Müller, Simone. "Toxic Commons: Toxic Global Inequality in the Age of the Anthropocene." *Environmental History,* vol. 26, 2021, pp. 444–50.

Nixon, Rob. *Slow Violence and the Environmentalism of the Poor.* Harvard UP, 2011.

Okafor-Yarwood, Ifesinachi, and Ibukun Jacob Adewumi. "Toxic Waste Dumping in the Global South as a Form of Environmental Racism: Evidence from the Gulf of Guinea." *Journal of African Studies,* vol. 79, no. 3, 2020, pp. 285–304.

Toxic Playground [Blybarnen]. Directed by Lars Edman and William Johansson Kalén, Laika Film & Television, 2010, https://www.imdb.com/title/tt1486586/?ref_=nm_knf_t2.

World Health Organization. Promotion of Chemical Safety Unit & International Programme on Chemical Safety. "Chemical Safety: Metabolism and Toxicity, Collected Lectures," 1993, https://apps.who.int/iris/handle/10665/59099.

4

Contesting Environmental Catastrophism

Queer Bodies, Afro-Indigenous Epistemologies, and the Crisis of Futurity

10

Culture, Inequality, and Queer Ecology in Rita Indiana's *La Mucama de Omicunlé*

Emily Baker

The Dominican writer Rita Indiana's 2015 novel, *La mucama de Omicunlé/ Tentacle,* is a not-so-futuristic tale of a society coping with a deadly virus while facing climate emergency and severe biodiversity loss. On a more micro level, characters deal with the day-to-day struggles of making money to survive, desires for sex changes, and the desire and/or pressure to produce art. With central characters as artists, electronic musicians, and gallery owner/ environmentalists, the novel exhibits a form of critical self-awareness vis-à-vis culture as these figures, and the reader by extension, are given multiple opportunities to see the future and to learn from the past, but they fail to do what it takes to avoid ecological disaster. In this chapter I argue that by pointing the finger at culture itself Indiana achieves the form of ecocritique advocated for by Timothy Morton: "a dialectical form of criticism that bends back upon itself," and, "as well as pointing, in a highly politicized way, to society, . . . points towards itself" (*Ecology,* 13). Furthermore, at the heart of the novel is attention to "race class and gender," which are "deeply intertwined with environmental issues" (13). Indeed, with reference to Morton, Jason W. Moore, James O'Connor, and others, I argue that an ecological Marxist lens is valuable for casting light upon the fundamental critique of racialized and gendered inequality, but that this Marxist perspective must also involve a queering of class. This ultimately points toward the need to bring together queer, Marxist, and ecological theory into a powerful analytical (polyamorous?) triad.[1]

Carlos Garrido Castellano's previous excellent study of this novel has examined the representation of culture as a function of "subalternity," yet without engaging in depth with the ecological or queer aspects, as my reading seeks to do. In an analysis that shares affinities with what follows, Samuel

Lagunas addresses the ecoapocalyptic events of the novel and asserts that *Tentacle* presents a "dura crítica al mesianismo y, al mismo tiempo, evidencia un fuerte desencanto con la literatura anterior, en específico con aquella vita épico-heroica que encarna *El reino de este mundo* (Alejo Carpentier, 1949) [hard critique of messianism and, at the same time, evidences a strong disillusionment with earlier literature, specifically with that epic-heroic biography that is embodied in Alejo Carpentier's 1949 *The Kingdom of this World*]" (192). Here Lagunas suggests an implicit cultural intertextual reference, usefully compared to other texts and forms of culture referenced explicitly by the novel, examined below. For their part, Sharae Deckard and Kerstin Oloff see the novel as a "Neo-Lovecraftian" rewriting of racist and ecophobic stories that manages instead to reveal "the legacies of colonialism and imperialism as constitutive of ecological violence in our current era of climate emergency" (3). Another study by Sebastián Figueroa and Lina Martínez-Hernández addresses intersections between queer, ecological, and anticapitalist dimensions of the novel, with disaster capitalism as its particular angle of approach.[2] Drawing upon, among others, Teresa de Lauretis, they interpret the failure to save the world as the deliberate frustration of our desire for resolution in fiction, questioning the power of language and "pervirtiendo cualquier esperanza depositada en 'lo humano' [perverting any hope deposited in 'the human']" (393). Building upon such readings, I place emphasis on the self-reflexive focus on art within the novel, which homes in on the potentialities and pitfalls of aesthetic production as, on the one hand, a means of changing the way we view the world and, on the other hand, a form of empty cultural capital, escapism, individualistic will to power, and at worst a perpetuator of the inequality that is central to the environmental crisis.

Davis and Todd demonstrate that narratives about the Anthropocene take an important stance with the dates they privilege in the account of the Anthropocene's origins, arguing for the acknowledgment of Indigenous scholars who see it as "the continuation of practices of dispossession and genocide, coupled with a literal transformation of the environment that have been at work for the last five hundred years" (761). Indeed, in *Decolonial Ecology: Thinking from the Caribbean World,* Malcom Ferdinand identifies and denounces "modernity's colonial and environmental double fracture," which "separates the colonial history of the world from its environmental history" (3). The temporal scope of the novel echoes the timeline of scholars such as Davis and Todd, Ferdinand, and Kyle Powys Whyte, who point to the conquest of the Americas and settler colonialism as decisive in the shifting of "ecological, cultural, and economic relationships with human societies and other non-human species" (Whyte 207). In one plot thread an artist called Argenis, from the 2000s, occupies—through a psychic time portal—the body of a buccaneer who is slaughtering cows and curing leathers to sell to British smugglers on the north coast of Hispaniola. This particular dimension of the novel, highlighting the exploitation of cheap natures (in this case, wild cow herds) under colonialism in the long sixteenth century evokes Jason W. Moore's emphasis upon the influence of capitalism as the principal ideology through which nature has been organized in modern times (172). This timeline and analysis underpins his argument for the use of the term "Capitalocene" to better suggest that it is not undifferentiated humanity that is responsible for the new geological age, but those that, historically and presently, have profited the most from capital accumulation (173).

The critique of capitalism is, as aforementioned studies have shown, central to the novel. In addition to scenes from the beginning of the seventeenth century, the novel is set in 2001 and the near future, post-2024, in the Dominican Republic, which has by then been plagued by "three disasters": a deadly and contagious virus, a tidal wave, and a 2024 seaquake that released Venezuelan biological weapons into the sea, provoking an extinction crisis (82). In what follows, I make reference to another ecological Marxist, James O'Connor, to describe the situation presented in the novel (and in reality) where, for the poor, survival and escapism from poverty is a barrier to ecological praxis, while for the rich, escapism, naive preservationism, and a failure to recognize their complicity in structural inequality is a barrier to environmental justice. In the final section, I address the ambiguous gender and sexual politics of the novel, in which a trans character is at the heart of the critique of self-actualization, binary thinking, and heteronormativity. Both Indiana and Morton suggest that a resistance to these would need to be at the heart of a (Marxist) queer ecology.

Acilde/Giorgio: Crossing Boundaries

The main character that transcends the three eras of the novel is Roque/Giorgio/Acilde, who simultaneously inhabits ~1606/2001/~2024, respectively. Born, notably, in the year of the financial crash, 2008, Acilde gives blow jobs to try and raise money for two dreams: one, to go on a cooking course and open a restaurant; and, two, to have a sex change, which by then is technologically possible by means of one prohibitively expensive and extremely painful injection that changes the sex of your body all at once. Following being raped by a gay Cuban man who thinks Acilde is a boy (emblematic of the vulnerability of their situation), Acilde is taken into employment by the personal guru of the despotic president, Said Bona, of the Dominican Republic, for whom the Cuban works. The Santería belief system provides the foundations for the rulers' decisions, and it turns out that Acilde is a "chosen one" in this scheme, destined to travel through time to the early 2000s and save humanity from the environmental disasters that we, and they, know to have unfolded.

Acilde has the sex change and travels back through a portal in the sea, emerging in 2001 in the man's body they wanted. Now Giorgio, he marries an upper-class hippie called Linda, who is on a mission to protect the island's coral reef. He fulfills his dream of setting up a restaurant and lives a happy life, throwing parties and commissioning art to fund-raise for Linda's coral reef project. In this state of self-fulfillment, he eschews the obligation to save the ocean and humanity. One of the final lines of the novel is "Tras hablar de rap y política, había despedido a Said sin decirle una palabra sobre su futuro. Podía sacrificarlo todo menos esta vida, la vida de Giorgio Menicucci, la compañia de su mujer, la galería, el laboratorio [After chatting about rap and politics, he'd said goodbye to Said without a word about his future. He could sacrifice everything except this life, Giorgio Menicucci's life, his wife's company, the gallery, the lab]" (132). Here, Giorgio, in a social encounter with the future president who presides over the environmental disaster to come, is content with mere superficial conversation and is not willing to give up his personal happiness in the name of the future of the livable planet.

Thus, at the heart of the novel, is a critique of self-actualization and escapism, particularly forms of escapism that are related to aesthetics. The aesthetic escapism (that the reader is complicit in) is particularly emblematized in the novel by art, culture (the discussion of rap and politics in the quotation above), and drugs and parties in the name of raising money for Linda's coral reef project. When they organize one such electronic music party, the DJ, Elizabeth (an upper-class character who tries to become an artist, without being very talented), remixes the voice of Jacques Cousteau into her set such that "las predicciones del explorador francés sobre el futuro submarino de la isla colgarían del silencio por unos segundos antes de que el bajo cayera otra vez, como un maremoto, sobre la pista de baile [the French explorer's predictions about the future of the island's marine life would hang in silence for a few seconds until the bass came down again, like a tsunami, over the dance floor]" (113). The metaphor mixes sea-based natural disaster (a tsunami) with the drug-fueled dance party, foreshadowing the disaster to come and structurally linking it to this upper-class hedonism.

The reference to electronic music may not be incidental. In *Ecology Without Nature,* Morton discusses the role of dance music relying as it often does on anonymous and machine-generated sounds as "ambience," which he describes as "a symptom of capitalist alienation" (87). The role of the DJ as remixologist obscures the labor that has gone into the original production of records, as well as the appropriation of political voices such as "Martin Luther King Jr., Ed Wood, and Gertrude Stein" (113) and Jacques Cousteau documentaries that she mixes into her sets, flattening them into "ambience" (89). Elizabeth's vocation as DJ is structurally tied in the novel to her experience of a Holy Week fertility ritual in which she has a revelation of "la extrema pobreza de los braceros haitianos, la boca trágica con la que este ceremonial antiquísimo se aferraba al presente, la permanencia de una esclavitud que se disfrazaba de oficio y el poder de una música que alojaba en el cuerpo humano a deidades capaces de tragarse al mundo [the extreme poverty suffered by Haitian workers, the tragic ties with which this ancient ceremony held on to the present, the permanency of a kind slavery that now dressed itself up as paid labor, and the power of music that lodged deities in human bodies, deities powerful enough to swallow the world]" (113). Yet her material response to this is simply to seek to find "el efecto bailable y misterioso de aquella fórmula mágica [danceable and mystical effects of that magic formula]" (113) for her own self-actualization and gain. Thus, in this first thread of Marxist critique, we can say that Giorgio, Linda, and their friends demonstrate how "human labor power and labor time get factored out of the process of value generation even though they are intrinsic to it" (Morton, *Ecology* 87–88). This applies also to visual art, as we see below.

Linda: The "Beautiful Soul"

Through Acilde's previous life, and other characters, we get to know that the upper and lower classes are structurally linked in a phenomenon whereby the people who need money do anything to get it, such as drug trafficking, sex work, and producing art, and the people who *have* money take drugs, pay for sex, and pay for art as escapism to try and mask their lack of fulfillment despite having money. Furthermore, instead of paying taxes or being involved in the true redistribution of wealth, the rich invent their own ways of "giving back." In the case of Giorgio, "El Sosúa Project . . . era una iniciativa cultural, artística y social con la que quería devolver algo al país que lo había hecho rico [The Sosúa project, as Giorgio called it, was a cultural, artistic, and social endeavor that he hoped would give something back to the country that had made him rich]" (29). As we see below, Linda's environmental savior complex is part of her own mode of "self-development and realization" (O'Connor 87). This fact comes through in an ironic comment when Linda is depressed about the fate of the coral and the world, and her brother comforts her, telling her "que no abandonara sus sueños, que el mundo necesitaba de más gente como ella y otras frases hechas de autoayuda que tienden a surtir efecto [not to abandon her dreams, that the world needed people like her, and other ready-made self-help phrases that had proven effective over time]" (92). The "self-help," to which the quotation refers, is a classic neoliberal tool, designed to soothe individuals into feeling that they are both relentlessly optimized and productive (to [green] capitalist ends) and simultaneously mellow and entitled to boundless "self-care."

All development is depicted in an ironic tone by Indiana. As Marx said in the *Manifesto of the Communist Party*, "The bourgeoisie cannot exist without constantly revolutionising the instruments of production, and thereby the relations of production, and with them the whole relations of society" (n.p.). Here, the infrastructure around the conservation of the coral reef is the next step in capitalism's productive renovation of itself. For example, the beach that Giorgio and Linda buy undergoes "eco-friendly construction," which means replacing a forest of brambles and *guasabaras* with a "casa de cemento y madera de estilo moderno, de un solo piso y con una enorme terraza de ladrillos que miraba hacia el mar [one-story concrete and wood house, with an enormous brick terrace that looked out to the sea]" (38). Far from actually being "eco-friendly," this project dewilds the surrounding landscape and plays into a modern aesthetic of "eco-friendliness" while actually being made out of cement, which contributes 8 percent of global carbon emissions annually ("Building the Modern World") and to marine ecotoxicity (Chen). It further enhances the ability of its inhabitants to enjoy the pleasure of their panoramic view of the sea. This represents a classically Romantic and objectifying approach to Nature, as Morton has it, which brutally reinforces the Nature/Culture binary.[3] It is part of a logic of "green consumerism," allowing people to be "both pro-capitalist and green, repeating the Romantic struggle between rebelling and selling out" (110). The modern take on Romantic consumerism is dubbed by Morton "reflexive consumerism," whereby consumerism represents "free-floating identity, or identity in process" (111). In other words, the purchase and design of the house are actions through which the Menicuccis' identities as "eco-friendly" are confirmed and performed (hypocritically).

Indeed, the way Giorgio's wife Linda views Nature is framed as both objectifying and dramatic:

Donde los demás veían paisaje, Linda Goldman veía desolación . . . Donde los demás veían un regalo de Dios para el disfrute del hombre, ella veía un ecosistema víctima de un ataque sistemático y criminal. Frente al arrecife de coral se sentía como un oncólogo ante el cuerpo de un paciente. Se sabía preparada para salvarlo, aunque también conocía al dedillo la desmedida capacidad del mal y su alcance.

[Where others saw scenery, Linda Goldman saw desolation . . . Where others saw a gift from God, given for the enjoyment of humankind, she saw an ecosystem fallen victim to a systematic and criminal attack. When she looked at the coral reef, she felt like an oncologist standing before her patient's body. She knew she could save it, although she also understood the disproportionate capacity of evil and its reach down to its finest detail.] (93)

Here, Linda falls into the trap of what Morton calls, after Hegel, "Beautiful Soul Syndrome," the Beautiful Soul being "a persona of the 'unhappy consciousness' that separates humanity and nature" (117). The clearly established dichotomy between what she sees and what others see, sets her up on the moral high ground, "fingers wagging . . . at modern society" (114), and she fails to acknowledge her own complicity in environmental destruction, as an upper-class person who takes material comforts for granted: "The beautiful soul maintains a critical position about everything except for its own position" (121). The use of the word *mal* (evil) is significant insofar as Morton uses the same term when he says, "the beautiful soul . . . cannot see that the evil it condemns is intrinsic to its existence, indeed, its very form as pure subjectivity, *is* this evil" (118). It is this abstracted gaze that reenforces the dangerous separation between Nature and Culture, epitomized by the fact that Linda sees herself as a savior (compared rhetorically to a cancer surgeon) and thus on a totally different plane to the "evil" that she must defeat.

Furthermore, it is the poor fishermen who embody the evil from her perspective: "Había días en que sentía que su compromiso era minúsculo frente, por ejemplo, al ancla de un pescador del pueblo, que en un minuto había arrancado un coral de cientos de años y aniquilado un valiosísimo espécimen y el hábitat de los peces que el mismo pescador necesitaba para subsistir [There were days when she felt her commitment was irrelevant, when confronted, for example, with a local fisherman's anchor that, in a single minute, had torn a reef hundreds of years old, destroying a valuable specimen and the fish habitat the very same fisherman needed to subsist]" (93). Here, in the literary conjunction, she is aware of the fact that fishermen rely on fish for their livelihood but does not see the tension whereby what she proposes would eventually exclude them from all coastal terrains in a "conservationist" logic that always rhetorically pits "nature" against "culture" or human activity (O'Connor 79–80). It is clear that the problem cannot be solved in isolation—by just privatizing the coastline and policing it to stop fishing without addressing how to provide livelihoods for those that this will displace. The overall challenge that inequality poses in these situations is emblematized in the following quotation: "Los guardias encargados de hacer cumplir las leyes ambientales en la Ensenada de Sosúa, eran los primeros en romperlas, echaban basura, pescaban con arpones y sacaban corales para venderlos, carentes de una preparación comprensiva y de sus sueldos adecuados [The guards charged with enforcing environmental laws in the Cove of Sosúa were the first to ignore them: throwing garbage, fishing with harpoons, and stealing coral to sell—they lacked a comprehensive education and adequate salaries]" (98–99). Here it is clear that regulation and laws will not be enough without a fundamental structural shift to provide adequate education and salaries for all. As O'Connor puts it:

> In today's capitalist economy, the linkages between particular cultures and configurations of nature, on the one side, and particular divisions of labour and technologies, on the other, are ... broken or long forgotten. In their place stands a commodified nature and culture of modernity, an ethos of self-development and realization (rather than that of "archaic" community). . . . The reaction is a plethora of culture and nature preservationist groups trying to protect or restore or remember this or that traditional cultural practice or landscape mostly abstracted from the prevailing methods of production and divisions of labor and types of commodities produced today. (87)

The Menicuccis are emblematic of a class that profit from their self-proclaimed environmental projects and foreclose class politics through their employment of economically precarious people. In the next section we see that the way they set up their preservationist fundraising business takes advantage of the surplus of artistic labor and the inequality in the Caribbean. Their terms of employment are quasi-feudal. Rather than monetary payment, the so-called offer you can't refuse (36) put forward by Giorgio to Argenis consists of "room and board for six months" (37). The gallery takes 40 percent of proceeds for all the work that is sold, and the artists are obliged to be solely represented by them. Thus, the Menicuccis literally take control of the artist's life and work. Having been thrown out of the house by his mother, Argenis is not in a position to turn the offer down.

Argenis and the Critique of Sovereignty

It is not just electronic music and "eco"-architecture that is criticized in the novel, but also fine art and design, by means of the character Argenis. Argenis is a divorced coke addict who works in a call center that offers psychic readings to callers from the United States, under the name "Psychic Goya." His name derives from his status as a graduate from a school of fine arts. He is skilled in copying the grand masters like Velásquez and Goya but when he attends, as a follow-up, the School of Design at Altos de Chavón, he realizes his Catholic themes and Renaissance perspective, however sophisticated, are out of sync with the time in which he is living. After hitting rock bottom and vomiting up a bottle of rum in the courtyard of the school where he has no friends, his professor takes him under her wing and gives him a crash course in contemporary culture and society: "*The Aesthetics of Disappearance, Society of the Spectacle, Mythologies, The Kingdom of This World, The Invention of Morel,* and *Naked Lunch*" (30). She tells him, "'Despierte, Goya, póngase las pilas, usted tiene una técnica impecable, pero no tiene nada que decir, mire a su alrededor, carajo, ¿usted cree que la cosa está para Angelitos?' [Wake up, Goya, get it together. You have impeccable technique but nothing to say. Look around you, damn it, do you think a bunch of little angels is what's needed here?']" (22). In the absence of a novelist or writer character in the novel, there is no explicit critique of written culture, and yet the fact that Argenis does not emerge enlightened from his contact with his mentor—and the books she recommends—appears to suggest that this aspect of culture may, too, fail to raise a sense of ecological consciousness. Through the discussion of art, however, we see a possible model for the successful transformative engagement of culture with society, which may be that which Indiana is trying to put into practice with her own work.

In stark contrast to Argenis, Goya, the artist who his colleagues compare him to, was known for exploring subject matter never seen before in the forms of painting, prints, and drawings (Tomlinson xxi). Goya was painting at a time of great political, social, and cultural shifts, and the novel suggests that we are at a similar crossroads of great upheaval. In the words of the novel itself, when Ivan, the manager of the art project to which Argenis pertains, gives a presentation on Goya's *Los caprichos,* he says, "Goya presenta una sátira subjectiva que no se amarra a una sola lectura, desestabiliza los paradigmas sociopolíticos de su tiempo a partir de personajes y situaciones que oscilan entre lo pintoresco local y lo mitológico universal [Goya presents a subjective satire that cannot be tied to any single reading, destabilizing the sociopolitical paradigms of his time with characters and situations that oscillate between the locally eccentric and the universally mythological]" (57). The chosen plate number 66 is described by the narrator in the following way: "Un cuerpo andógeno y retorcido sostenía el palo de una escoba voladora sobre su cabeza u ocultaba tras de sí un cuerpo de formas femininas más evidentes que también se asía al palo y desplegaba alas de murciélago que facilitaban el vuelo mágico [A twisted androgynous body held a flying broomstick above his head, obscuring the more feminine figure behind him, who also held on to the broomstick and sprouted bat wings to facilitate the magic ride]" (57). The image said to achieve the "destabilization of sociopolitical paradigms," therefore, is one that depicts androgyny and, expressed in another way, posthuman hybridity. Yet these are things that Argenis flatly rejects.

Argenis is misogynistic, homophobic, racist and "la pasión ambientalista le importaba un bledo [could not give three fucks about their environmentalist passions]" (40). When Giorgio proposes to hire him, he gives him a friendly hug but, "pero a Argenis la cercanía física de un hombre le activaba un avispero bajo la piel [any kind of physical closeness with another man made him break out in hives]" (36). Later we learn "al pintor, el contacto de otros hombres frente a él también lo hacía sentir incómodo [contact between two men, and so near to him, always made Argenis uncomfortable]" (39). While this homosocial contact is disturbing to him, Argenis spends most of his time, as he is inducted to the gallery and sanctuary, having explicit sexual fantasies about Linda. She makes an innocent comment to him, inviting him to go snorkeling as part of the group, and he thinks to himself, "Me la saqué, no se lo va a dar al Prieto e Iván debe ser maricón, quiere que yo se lo meta [She's mine! She's not gonna give it away to the nigger and Iván must be a faggot. She wants me to fuck her]" (the racist, sexist, and homophobic language is especially evident in the English translation; 42). As they are all diving together under the sea, Argenis follows Linda through a hole in the rock, trying to get a look at her ass, and gets badly stung by anemones, having to be rescued and given adrenaline (Gaia's punishment?). In the following statement we see his racism and critique of the Cuban Revolution: "Todo el mundo los ama, por la maldita revolución.... ¿Qué es lo que me va a enseñar este pelafustán? Si allá la gente da el culo por una pasta de dientes [Everybody loves them because of their goddamned revolution. . . . What's this layabout going to show me? I mean, people back in Cuba give their asses for a tube of toothpaste]" (33). This personality places Argenis in the position of fierce defender of binaries and purities that are the opposite of the hybridity, fluidity, and plurality of the "queer" mode. As Morton warns, anything that upholds a subject/object dichotomy or "inside-outside manifold," evident in the way that Argenis objectifies Linda, Malagueta, and Iván for their gender, race, and sexuality respectively, is akin to a type of environmentalism that fetishizes purity and upholds the false Nature/Culture binary. The need to bring the queer and the ecological together, for Morton, derives from the fact that they are both defined by the values of nonessentialism and a lack of sovereignty: "Nothing exists independently, and nothing comes from nothing" ("Queer Ecology" 275). Both require "a multiplication of differences at as many levels and on as many scales as possible" (275). Argenis, in addition to rejecting others physically in a sort of impulse for sovereignty, is also defined by his production of sameness by means of his art, which merely copies the work of grand masters rather than creating anything new. Like Giorgio, Argenis experiences a form of time travel that might bring the opportunity to transform him. However, like Giorgio, it turns out not to be truly meaningful.

Judging by the experience of Acilde/Giorgio in traveling through time via a portal in the sea, we must surmise that Argenis, in swimming through the rock circle following Linda, must also have been psychically transported into another time and body through that process. As Argenis tries to make sense of his hallucinations, he realizes he has been taken to the period after 1606, when General Osario organized the depopulation of large sections of the Hispaniola coast in order to stop smuggling.

There are a number of structural parallels that are made between the two eras that might shed light on Indiana's purpose in revisiting the seventeenth century. Argenis, having been rescued from the sea by local smugglers, has to engage in curing leathers and smoking meats of wild cattle in exchange for food and community. As Juan José Ponce Vázquez outlines, "Hides were in high demand for the manufacturing of a myriad of goods, from clothing and boots to weapons, tools, saddles and bridles" (47). While Argenis gets used to the sights and smells of catching wild cows, skinning them, drying out their hides and meat, and eating the jerky, Giorgio, in the present, cooks massive beef steaks on his grill. Here (as with his house made of concrete) he conveniently ignores the environmental impact that red meat, particularly beef, is known to have. This juxtaposition of Giorgio's steak with Argenis's cured meats stimulates a reflection on the evolution from this exploitation of "cheap natures" (Moore)—in other words, opportunistic killing of cattle from wild herds to the types of industrialized farming that will have produced the modern-day meat they are eating.

Another more indirect parallel between the two eras relates to the phenomenon of smuggling as a means of survival for impoverished locals in Hispaniola and other parts of the Caribbean. In the same way as Acilde is forced to give blow jobs for money, and Argenis works at a call center, the locals did what they could to ensure there was bread on the table and to keep themselves clothed in accordance with their social expectations. In his work, Ponce Vázquez identifies a "moral economy of smuggling" whereby "in Hispaniola . . . this social ethos also cut across race and class, unifying the great majority of society into a communal pact that normalized their practice of and benefits from contraband trade, while maintaining a deeply hierarchical society in which enslaved and free people of color, either willingly or forcedly, participated in the trade often as cogs within a smuggling machinery controlled by white elites" (2). In the vein of a postcolonial critique, it is clear that the social relations established in this colonial era have provided the foundation for what comes subsequently. The Menicuccis' project represents a microcosm of this society in which, on the one hand, there is interracial and class cooperation (Argenis, Iván, and Malagueta working with Linda and Giorgio) but the terms under which they work ensures the ongoing existence of a "deeply hierarchical society" (Ponce Vázquez 2). While they are not engaged in the same activity of smuggling, there are parallels here, too, insofar as both are privatized activities that bypass the state. In the seventeenth century the crown was struggling to maintain control in this jurisdiction, while in the 2000s neoliberal capitalism has ensured that the state has all but withdrawn, and the market and private enterprise (including environmental preservationism) prevail. Eventually, however, a Spanish galleon will force the smugglers into hiding, and the "three disasters" of 2024 will render Linda's coral reef preservation useless.

In the process of traveling through time, Argenis appears to undergo a transformation that sees his homophobia diminish and his capacity for producing original art increase. His desire for isolation and aversion to touch seems to fade, as he has to cooperate with the buccaneers in his seventeenth-century world to survive. He even receives a blow job from Roque, the leader of the buccaneer group, in a sharp reversal of the aversion to physical contact between men that he expressed previously. Unbeknownst to him, this is actually Giorgio who has been transported further back in time. This affection between them translates back into the present, when Argenis fantasizes about hugging Giorgio as he witnesses an argument between him and Linda, "dándole la oportunidad que Argenis ansiaba de aconsejarlo, de devolverle agradecido su amistad, de rodearlo con un abrazo en el que su pecho tocaría el de su mecenas finalmente [giving him the opportunity he so wanted to counsel him, to gratefully be a friend to him, to put his arm around him and let his chest finally touch his patron's]" (95). It is as if the experience of being a buccaneer has released a latent homosexuality that Argenis may have repressed following a violent shaming he experienced after kissing his father on the lips as a young boy. His transformation is, however, not fully for the better, as he now appears to feel a violent hatred toward Linda: "Fantaseó con violarla y estrangularla, luego con machacarle la cabeza con el bate de béisbol de aluminio que tenía Malagueta en su taller [He fantasized about raping and strangling her, then about beating her head in with the aluminum baseball bat Malagueta kept in his studio]" (95). His detached objectification of women has translated into outright fantasies of violence and murder. This violence ultimately manifests not toward Linda herself but her beloved dog, Billy, whom Argenis deliberately and callously poisons in an act of spite.

In addition to his romantic desire for Giorgio/Roque, true creativity is unleashed as he experiences a shift in perspective. In his seventeenth-century world, Argenis uses a press and wooden blocks to make engravings recording the lives and activities of his buccaneer companions, mirroring his artistic enterprise in the 2000s. The complicity of his art with the violence of the killing of the wild herds is starkly represented by means of the fact that he uses the clotted cow blood to make the prints, in the absence of paint.[4] In the present, his creativity seems associated with the desire for Giorgio: "Hacía siglos que no miraba a nadie como ahora hacía con Giorgio [It had been a very long time since he'd looked at anyone the way he was looking at Giorgio now]" (60). He paints the portrait of Giorgio in the perfect evening light on the terrace, juxtaposed with his own distorted and deformed face as he surfaced from the sea with the anemone stings. "La cara anaranjada era altiva y hermosa; parecía dar una orden, que el monstruo deforme, a juzgar por la inclinación de la cabeza, acataría diligente [The face lit by the candle was haughty and beautiful; it seemed to giving an order with which the deformed monster, given the inclination of its head, would comply]" (61). In this painting, as in the world, the "good" and beautiful is structurally linked to, and balanced by, the monstrous. In the buccaneer world, Argenis is eventually forced to bury his prints when fleeing from the punitive Spanish galleon, only for them to be discovered in the present as "treasure" while a new area of beach is being dug up for Linda's marine laboratory. This crosscutting of past and present precipitates what is interpreted as a psychotic episode whereby Argenis claims the treasure is his despite the fact that it has been found on the Menicuccis' land. In this moment of convergence between past and present, Roque (Giorgio in the past) takes the opportunity to shoot Argenis, which Roque/Giorgio frames as an act of revenge for the murder of his dog. In the present, Argenis's outburst consolidates his social ostracization, which he had already been working toward with his solitary and bitter personality.

Toward a Queer Marxist Ecology

Through *La mucama de Omicunlé* Indiana achieves a critique of many forms of so-called environmental politics and culture that is, as Morton demands, "critical and self-critical" (*Ecology* 13). As a singer/songwriter/DJ as well as attendee of Altos de Chavón School of Design, like Argenis, it is clear that there is a little something of Indiana in many of the characters depicted. As Morton says, "The only ethical option is to muck in," and they conclude that "*environment is theory* not as answer to a question . . . but as question, and question mark, as *in question,* questioning-ness" (175). The novel raises many questions as it wades into the complex realities of escapism, self-actualization, inequality, racism, homophobia, and the upholding of dualisms of sex, class, and, ultimately, Nature vs. Culture, mediated as we see by capital. No character can escape their complicity and thus neither can the author or reader, by extension.

The sexual and gender politics of the novel are complicated, to say the least. One of the key questions has to be: Why does Indiana place the burden of saving humanity from ecological destruction, and the critique of self-actualization, on a trans character (Acilde/Giorgio)? Could it not have just been a cis-heterosexual male at the center of this failure? Or is the problem that this is effectively what Giorgio becomes? As Lagunas acknowledges, "La propuesta ética implícita de la novela: solamente abriendo una puerta en los muros de la temporalidad lineal y la corporalidad cisheteronormativa es posible articular las resistencias del futuro [The implicit ethical proposal of the novel [is that] only by opening a door in the walls of linear time and cis-heteronormative corporality is it possible to articulate resistance to the future]" (190). In more ways than one, Giorgio ultimately represents the opposite of the queerness that Morton suggests we need to affirm when he reminds us that ecology is derived from biology, with all of its nonessentialist aspects, and that queer theory is a nonessentialist view of gender and sexuality; therefore, "the two domains intersect" ("Queer Ecology" 275). It turns out that each character examined in this chapter fails to adhere to Morton's definition of queer ecology in specific ways. Giorgio, who never tells Linda about his previous life as a woman, might be seen to embody "heterosexist gender performance" (Morton building upon Judith Butler's case for queer ecology), which "produces a metaphysical manifold that separates 'inside' from 'outside'" akin, as Morton posits, to the process by which the environment is figured as "a metaphysical, closed system": Nature with a capital N (274). Linda emblematizes this reification of Nature through her "beautiful soul syndrome," which subordinates the protection of Nature to a function of her own self-actualization. Moreover, it is clearly the kind of project that seeks to "repress the abject" (pollution/dirt/toxicity), domesticate (via the "eco-friendly" construction) and thus aestheticize Nature further (274). This is the opposite of a queer ecology that "must visualize the unbeautiful, the untold . . . the unsplendid" (280). Indeed, Morton points out that the type of environmentalism that fetishizes the "pristine" and represses the abject is precisely akin to kinds of racism and homophobia (displayed by Argenis) that also fetishize purity in different ways. While Argenis eventually accepts a form of homosexual desire, this is coupled with a desire to eliminate Linda and/or her dog. Thus, his murderous actions as well as his artistic production of sameness and his desire for sovereignty represent the opposite of the "intimacy" with other beings, human and nonhuman or "strange strangers" that for Morton (derived from Derrida), is the cornerstone of queer ecology (277).

If Marxism is brought into the queer ecological equation, the characters fail on this account as well. Giorgio not only transitions from female to male, but also from lower-class sex worker to upper-class restaurant and gallery owner. Instead of celebrating the forms of hybridity that he embodies in both cases, he crosses to the other side of the binary. He certainly does not use his new class position to advocate for the oppressed, but instead exploits his workers. Likewise, Linda's preservationism has little concern for the fishermen that rely on fish stocks to subsist. Argenis, too, is unable to buy into a cross-class project (albeit flawed, as we have seen) to help preserve the biodiversity of the region, and his actions are always motivated by capital gain.

To incorporate elements of Marxism into the theoretical frame, then, we need to foreground a nonessentialist vision of class within our analysis of capitalist-ecological relations. This conforms to what O'Connor alludes to when he says that ecologically speaking the struggles pertaining to production conditions are "class issues (and *more than* class issues)" (14). Struggles for meaningful ecological futures have to cut across traditional class boundaries if they are to be successful. In a notable moment Giorgio says to Argenis, "'Tenemos que proteger el mar o si no,' dijo, y con la mano hizo una pistola que puso sobre la cabeza de Argenis y disparó ['We have to protect the sea or . . .' he said making a gun with his hand and pointing it at Argenis's head before shooting]" (40). It is significant that Giorgio, the wealthy entrepreneur, puts the gun to the head of Argenis, the struggling artist, not his own. The rich will always have forms of protection and mobility accessible to them that the poor do not when displacement due to environmental disaster begins to affect more and more areas.

It is clear from Indiana's novel that we need to work toward what O'Connor describes as "a class politics that addresses local environmental issues and political identity as interconnected with problems of unemployment, low wages, homelessness, poverty, inequality and social decay" (88). Moore advocates for the use of the term "Capitalocene" instead of "Anthropocene" because, with reference to the latter, "inequality, commodification, imperialism, patriarchy, racial formations, and much more, have been largely removed from consideration" (170). It is not the "anthropos," an undifferentiated mass of humanity, that bears the burden of the transition to an era in which human activity is registered in the geological fabric of the earth, but the portion of humanity most heavily driven by a capitalist logic, and with the most capital to play with. In *La mucama de Omicunlé* we see that capitalism mediates the relationship between humans and Nature regardless of whether characters are rich, poor, well intentioned vis-à-vis the environment or not; but the rich have much more power to dewild, travel, pollute, and consume. This destructive relationship is further accentuated by an aestheticization of nature, whether that be selling coral reefs to the tourist industry to supplement a poor wage or building a terrace to better admire a sea view.

At the end of the novel, Giorgio presides over the death of his seventeenth-century self and the suicide of his self of the 2020s. This is the opposite of O'Connor's appeal that "the point . . . is not only to understand the past, but to change the future" (88). In its critique of Giorgio's choices, the novel implicitly celebrates androgyny, hybridity, and sincere meaningful action, divorced from pretense, identity-affirmation, and escapism; it acknowledges that the pollution, dirt, and messiness we have created is here to stay, and it conveys that to prevent further environmental catastrophe we have to embrace a truly multiclass, multiracial, multigender, anticapitalist movement, right now.

Notes

1 Poststructuralist forms of Marxism have already deessentialized class, and this position is in line with Morton's ecocritique that "fearlessly employs the ideas of deconstruction in the service of ecology and is 'similar to queer theory' in not being 'afraid of nonidentity'" (13). I demonstrate that Indiana is concerned with the failure of characters to experience cross-class solidarity and their inclinations to settle into identity positions (whether those be defined by gender, sexuality, class, "ecofriendliness," or a combination).
2 See also Estévez Ballestero for an account of the connections between "disaster" and "body" in the novel.
3 I use a capitalized version of "Nature" in the same way as Morton, to defamiliarize it.
4 In her pleasingly situated reading, Perkins analyses these references as a function of the novel's production of an "alternative material history of the Caribbean" (51).

Works Cited

"Building the Modern World: Concrete and Our Environment." Science Museum, 2021, https://www.sciencemuseum.org.uk/objects-and-stories/everyday-wonders/building-modern-world-concrete-and-our-environment. Accessed 21 March 2022.

Chen, C., et al. "Environmental Impact of Cement Production: Detail of the Different Processes and Cement Plant Variability Evaluation." *Journal of Cleaner Production*, vol. 18, no. 5, 2010, pp. 478–85.

Davis, Heather, and Zoe Todd. "On the Importance of a Date, or, Decolonizing the Anthropocene." *ACME: An International Journal for Critical Geographies*, vol. 16, no. 4, Dec. 2017, pp. 761–80.

Deckard, Sharae, and Kerstin Oloff. "'The One Who Comes from the Sea': Marine Crisis and the New Oceanic Weird in Rita Indiana's *La Mucama De Omicunlé* (2015)." *Humanities*, vol. 9, no. 3, 86, 2020, pp. 1–14.

Estévez Ballestero, Melania Ayelén. "Cuerpos del Desastre: Mutantes, Transformistas y (A) normales." *Caracol*, no. 18, 2019, pp. 83–100.

Ferdinand, Malcom. *Decolonial Ecology: Thinking from the Caribbean World*. Polity, 2022.

Figueroa, Sebastián, and Lina Martínez-Hernández. "Ecologías queer caribeñas y capitalismo del desastre en *La mucama de Omicunlé* (2015) de Rita Indiana." *Tekoporá. Revista Latinoamericana De Humanidades Ambientales Y Estudios Territoriales*, vol. 3, no. 1, 2021, pp. 391–408.

Garrido Castellano, Carlos. "'La elocuencia que su entrenamiento como artista plástico le permitía.' Subalternidad, cultura e instituciones en *La Mucama de Omicunlé* de Rita Indiana Hernández." *Hispanic Research Journal*, vol. 18, no. 4, 2017, pp. 352–64.

Indiana, Rita. *La mucama de Omicunlé*. Editorial Periférica, 2015.

Lagunas, Samuel. "Imaginarios teológicos en la ciencia ficción Latinoamericana reciente. *Zombie* de Mike Wilson y *La mucama de Omicunlé* de Rita Indiana." *De Raíz Diversa*, vol. 8, no. 15, 2021, pp. 171–95.

Marx, Karl. "Manifesto of the Communist Party." 1848. *Marx/Engel's Selected Works, Vol., 1*. Progress, 1969, marxists.org. Accessed 25 February 2022.

Moore, Jason W. *Capitalism in the Web of Life: Ecology and the Accumulation of Capital*. Verso, 2015.

Morton, Timothy. *Ecology Without Nature: Rethinking Environmental Aesthetics*. Harvard UP, 2007.

Morton, Timothy. "Guest Column: Queer Ecology." *PMLA*, vol. 125, no. 2, 2010, pp. 273–82.

O'Connor, James. *Natural Causes: Essays in Ecological Marxism*. Guildford, 1998.

Perkins, Alexandra Gonzenbach. "Queer Materiality, Contestatory Histories, and Disperse Bodies in *La Mucama de Omicunlé*." *Journal of Latin American Cultural Studies: Travesía*, vol. 30, no. 1, 2021, pp. 47–60.

Ponce Vázquez, Juan José. *Islanders and Empire: Smuggling and Political Defiance in Hispaniola, 1580–1690*. Cambridge UP, 2020.

Tomlinson, Janis A. *Goya: A Portrait of the Artist*. Princeton UP, 2020.

Whyte, Kyle Powys. "Our Ancestors' Dystopia Now: Indigenous Conservation and the Anthropocene." *The Routledge Companion to the Environmental Humanities* by Ursula Heise, Jon Christensen, and Michelle Niemann, Routledge, 2017, pp. 222–31.

11

Toward a Cosmopolitics of the Image

Notes for a Possible Ecology of Cinematic Practices

SEBASTIAN WIEDEMANN

This chapter was originally written in the Global South, in Portuguese and Spanish, deeply affected by Amerindian forces. And I have been willing to circulate it in this context and in this supposedly lingua franca by an invitation to contribute to decolonizing a certain way of inhabiting the production of thought in the university. Diverse ways of thinking also demand diverse ways of reading and listening, not always aspiring to the transparency of a common language, codes, and media. That is to say, to open oneself to the vertigo of the opacities that implies welcoming a pluriversity rather than a university. The decolonization of thought cannot be just a topic to be treated; it must be, at the same time, a form that introduces the aberrant in the orthodoxy and hegemony of Western thought with its lingua franca, English. Form and content as phases of the same deviant movement in the face of the norm and in search of an authentic appeal and challenge for thought. Having said this, and inhabiting academic English from a wild perspective that lets the original Portuguese and Spanish still breathe in a clandestine way (and to let them breathe is also to let breathe a borderline thinking of the edges between hemispheres and worlds), I can argue that, starting from a performative disposition, this chapter proposes to be a gesture of philosophical and audiovisual experimentation. That is, a gesture that introduces a mixed plane of thought between audiovisualities and conceptualities from which, in an immanent way, emerges the notion of cosmopolitics of the image, as an emanation of an ecology of practices in the act that nourishes itself from the encounter and becoming with potencies of thought of the difference, of matter expressiveness, and of extramodern peoples like the Yanomami of the Amazon rainforest. Ultimately, the attempt at an appeal and a challenge for thought is a decolonizing gesture of the image through the proliferation of cinematic modes of experience.

*

We can say that thinking is always thinking by other means or media. Media, however, understood here as sliding and passing surfaces, are not just materialities like film and paper, but also materialities like other peoples' cosmogenetic potencies. This is the issue that I want to explore in this chapter by addressing two films, *Abyss* (2012) and *Xapirimuu* (2016), which I have produced in the encounter with various cosmogenetic potencies—that is, the molecular potencies of color, the Yanomami potencies of thought, and the philosophy of difference.

I would like to start by proposing a formula that seems effective to express the demands of the gesture that I want to unfold here: *to make exist and not to judge, being worthy of the otherness*. This formulation undoubtedly resonates writings by Gilles Deleuze (*Lógica do Sentido; Crítica e clínica*) and Isabelle Stengers toward a speculative empiricism (Savransky). This cothinking also moves us toward anthropology and the way it approaches Amerindian worlds. I propose *thinking-with* from the perspective of a filmmaker, or, better yet, a practitioner of cinematic modes of experience. This thinking through multiplicities and partial and transversal connections (Strathern) gives consistency to what Stengers calls an ecology of practices ("Introductory Notes on an Ecology of Practices").

Like Stengers, I cannot deny that I am an heir of the Enlightenment and the monotheistic creed. The cinematic practice is neither neutral nor innocent. Cinema's infancy is tiny if we think of how long it took for a clear distinction to be made between magic, alchemy, and chemistry. Cinema is a product of modern life, of modern thought; it was already born in a major key, to use Stengers's words (*In Catastrophic Times*). A major key, a premature majority that confines the potential for experimentation and creation of the practice while subjecting it to the logic of representation and the market. In the primacy of naturalism, there are two options: either the world becomes an image, or we are given an image of the world. For the practices I claim within the formula/proposition presented before, and that I want to unfold here, what should emerge is the genetic potency of a world without an image. This is a world that is always to be made, a pluriverse, if we follow William James. Certainly, the cinematic apparatus reproduces what Alfred North Whitehead called the bifurcation of nature (*The Concept of Nature*), except that for a nature, for a univocal world, there is a multiculturalism instead of a multinaturalism (Viveiros de Castro, *Metafísicas canibais*). It is then necessary to step outside of cinema by means of cinema itself. A decolonization of

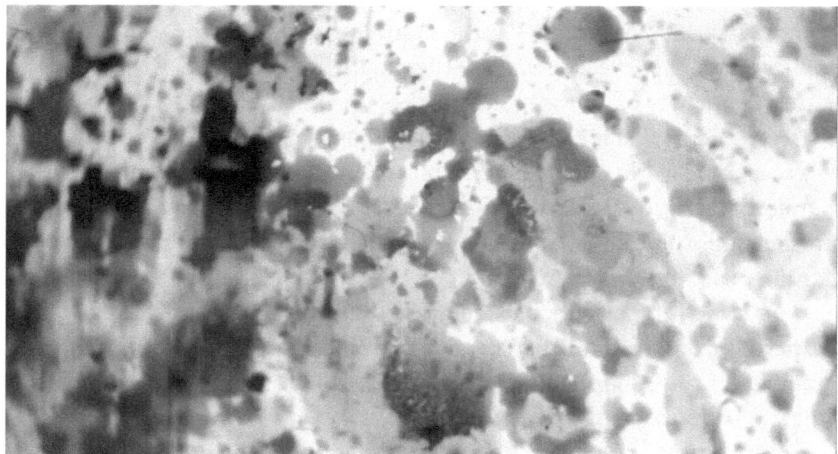
Figure 11.1. Screenshot of film *Abyss* by Sebastian Wiedemann.

cinema is imminent if we want it to stop judging, reproducing, representing; if we want cinema to become a means for an effective appeal and challenge for thought.

We speak of the emergence of a cinema before knowing Cinema, of a minor cinema, porous and open to contagion and unthought alliances, where the instability and vulnerability of becomings are embraced in the absence of foundation, of essence, or of a priori forms. We mean a cinema that takes risks, that experiments. An experimental cinema that gives vent to the cosmos, to connections through differences, and, as a consequence, claims to be cosmopolitical. A cinema in which the concern is no longer the reproduction of a world and thus the reproduction of a moral. A cinema that, as a pure mode of experience (James), immanent as life itself, can only be called a cinema of process (Whitehead, *Process and Reality*).

In this cinema of process, (cinematic) modes of experience proliferate, with two propensities. On one hand, there is the interval, which, as we will see later, is not far from the gesture of being worthy of the otherness; and, on the other hand, there is the issue of how, in the midst of a cosmogenetic process of proliferation of heterogeneities, the visualities and sonorities present themselves as attractors of movement and as movement of thought. In other words, the series is and should be open, including the most diverse modes of experience, as well as the contagion between divergent means, surfaces, and practices. However, the visualities and sonorities are the propulsive vectors of thought, *abyssing* it in the interval, from where a novelty, a disparity, or a differential may emerge. The a-signifying quality of the visualities and sonorities tends to create greater vertigo in the diagram, and, with it, there is a greater potency of falling, which makes the image always that which is to come. A cinema that is concerned with the constant and unending unfolding of a pluralistic world—with the variability, transmutability, and multirelation (Orlandi) of perceptual matrices, and whose birth hatches from the intervals. This metamorphic cinema can certainly be independent of film. As it is happening in act in this very instant, it can become film by other means. It can also be its writing and its conceptualization practice. In this sense, could we perhaps, and from the perspective that we are defending, affirm that the practices of the Yanomami shamans in relation to the *xapiripë* are already cinematic modes of experience, as a kind of somatic and embodied cinema?

Before testing the efficacy of these speculations, I suggest that we face the vertigo of a first interval with my film *Abyss* (2012).[1]

In order to decolonize cinema, its expressive potencies should be driven to the highest level of singularity. No cinema can be constituted as an ecology of practices if it stands as general and generic, intending to reproduce a universalizing will. Its aprioristic fabric should be dismantled. Cinematic potency cannot be reduced to figuration, representation, and its cause-and-effect rationale. We cannot even presuppose that it needs a camera to be effected, or that it is the subsidiary of a will to analogize with reality and with vices inherited from literature and theater. Cinema can occur as a cinema-paper, like the very thing you are now hearing/reading, but also as a tactile gesture of painting over the materiality of the film itself, such as in *Abyss*. This kind of cinema is decolonized from any transcendent will and moves on *to make body with*. It refuses to abandon pure experience (James), so, as the practice emerges, so does the practitioner. What is at stake, for instance, when *making body with* film and painting a photogram after a photogram, is how to lengthen the duration of the occasion where, mutually, practice and practitioner are integrated into the cosmogenetic process, in the process of giving birth to worlds. This process may integrate as many components as ecology can give consistency to and is able to sustain in the plane of experience, all the while knowing it will never achieve self-sufficiency. It is always in a to-be state, much like the image is always to-come. How many chromatic and molecular varieties can film withstand? How many different, unstable worlds can the practice of handmade cinema, using a Pollock-inspired *dripping* technique, spawn, and proliferate? For a cinema of process, the intensive qualities will always be of more interest than the extensive quantities, since it is in the intensive plane that the components will necessarily be in their highest level of activation—of experimentation, experience, and responsiveness.

The world is being made into pluriverse, and so are we along with it. Folded in experience, everything is in movement, and also in a minor key. This cinematic thought in action, which never ceases to take place, erases hierarchies, and, in their absence, the challenge begins. The ecology of practices in its porosity and low adherence is a tool to think what is going on (Stengers, "Introductory Notes on an Ecology of Practices") while it is going on and where, in the vertigo of the new, the interval calls for it; it, the tool, cannot be fixed on a habit of recognitions. Each time around, the ecology is another; each time around, the assemblages are different. Agencies never proliferate the same way, for what is proliferating is a connective and rhizomatic body of affects among different configurations of divergent worlds without a common aprioristic core. Ink and celluloid do not presuppose a relation. Each time around, the plane of cinematic experience is remodulated as a critical area of complexity and heterogeneity, always with a local foothold, but nonlocatable, which means it will not be crystallized or sedimented as image. Rather, it will be a source for this image that always presents itself as a potency of futurity, of what is to come.

In this flux of experience, there is no film, although one is affirmed. There is above all a situation-film that does not tell us what to think, that does not claim to be a watchword—it does not judge, it brings to existence—because what is at stake as a decision is to give the situation the power to make us think (Stengers, "Introductory Notes on an Ecology of Practices"). In the ecology of practices, instead of wielding power, potency is distributed. There is no film in it, but the situation-film unfolds the virtuality of an infinity of films that can precipitate unexpectedly, which can be actualized as a colorful spray against celluloid. That is, this cinema of process does not manage nor control the power of acting and being affected, but it bestows power to the situation, embracing turbulence and the unpredictability of encounters. As the becoming catalyst of the differential, it is allied with chance, with the cosmos, right there where the situation becomes a matter of particular interest and concern—in other words, right there where it makes us think instead of recognize (Stengers, "Introductory Notes on an Ecology of Practices"). As Deleuze puts it, it happens right there where it makes us think in between, where there is no place for anchoring definitions or for an ideal horizon. *Abyss* can begin and end at any point, and no presupposition or theory would give us the power to pluck out any component that composes its particular medium. In other words, while thinking from within, while making body with an ecology of practices, we cannot afford the luxury of abandoning the particular, the singular, and the internal relations by name or direction to something that we would be able to recognize or identify in virtue of having a prior image of that something. This cinema of process does not want to have images; it, rather, wants to move, as it is always sniffing out an image that is endlessly slippery. So, it learns to slip continuously, to *make body with* immanence, and to disregard any abstraction that pulls it away from its medium. As an ecology, it is always an ethoecology, a tangled complex of processes of intensive mutual affecting. Or, to say it another way, we do not know what a practice can do or be, we do not know what it is capable of becoming (Spinoza).

Thinking from the standpoint of radical empiricism, it is always crucial to ask: What is reality capable of? (Savransky). It is unpredictable, much like it was impossible to predict that the emergence of *Abyss* would cause, in its turn, the emergence of this film-paper-writing folder in this ecology of practices. We cannot predict which ethos the practices will secrete, because, when thinking in a minor key, what is in contention is an ethics of encounters that, at each turn, is affirmed in a singular instance. That is why Deleuze (*La imagen-tiempo*) says of experimental cinema that it rather searches, instead of attempting to find. And I will add: it always searches without horizon.

This ecology of cinematic practices must refrain from any will of truth. There is no truth in the occurrence of thought, where a proposition is unfolded and acquires expression. We can only ask ourselves about the efficacy of the way it affects and engenders thought, about the efficacy of the way it is composed actively and potently with the medium of the practice which it is a part of and to which it contributes to give consistency. Hence, all ecology of practices is a pragmatic gesture that favors precarity and turbulence over the protection of a general, transcendental reason that offers it shelter but, in return, atrophies its delirium and possibility of venturing, of bringing to existence. And not only that is at stake; as when intimacy is established with a practice, we the practitioners belong to it, we cocreate ourselves, we coproduce each other, and, in some measure, that imposes some obligations.

By affirming myself as a filmmaker, I have obligations, as the movement of ecology also depends on me and the movement in me depends on the ecology; this is an interconnection that cannot afford the luxury of being abandoned in favor of the comfort of truth and the abstractions that sustain it. It is about taking part in a decision without a decision-maker; it is feeling in the body, in the *making body with*, that, in the absence of a foundation, the body of the bodies of ecology is sustained, as I belong to it and am capable of

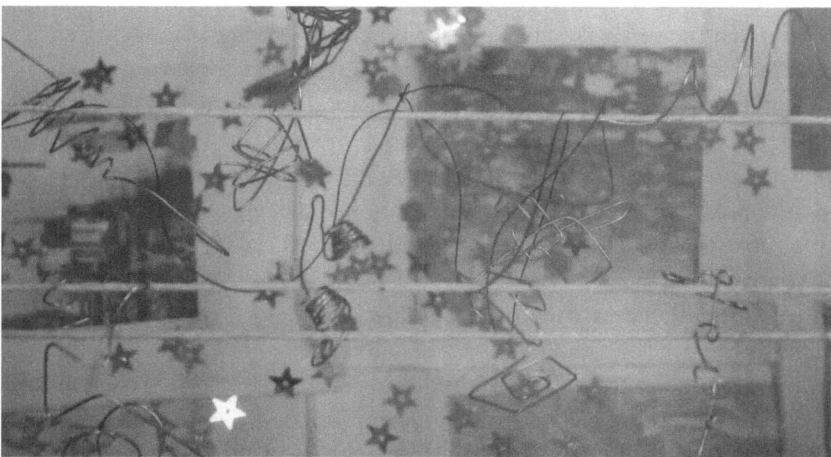

Figure 11.2. Screenshot of film *Xapirimuu* by Sebastian Wiedemann.

doing that which I would not have been able to do otherwise. I do not make the cinema that I would like, but the one that I can. The one that I need, the one that I am called upon to make. Ecology secretes me, much like it secretes this text that you are reading. Much like, in its turn, the affects that it activates in you make it so that, even provisionally, you also belong to it. It is a risk in itself. And every encounter is a risk, and without risks there is no thought. This risk speaks of a fragile potential unit that emerges as a common cause of this abyss that is the interval. Thus, if the encounter occurs, as Stengers puts it, it must be celebrated as a "cosmic event" (*In Catastrophic Times*). The search for these minimal events, always involuntary, like the germ of an *appeal and challenge for* thought, is what moves the ecology of practices, as an experimental being-together of pragmatic and dynamic learnings about what works and how it works. These learnings make this search a cosmopolitics in the unfolding and affirmation of its potency as a radical constructivism—that is, *about bringing into existence and not judging, being worthy of otherness.*

I now propose a second interval with my film *Xapirimuu* (2016).[2]

Otherness, as Deleuze (*Lógica do Sentido*) reminds us, is the expression of a possible world. In the approach we are claiming about the singularity of this ecology of cinematic practices, we will say that the otherness is a fold in the interval, as a differential potential in constant dynamism. Otherness is understood as this potential to alter worlds, or to make a pluriverse of the world. It is a potency of difference, divergence, and proliferation as a relational condition between disparities that take turns in the possibility of saying that the world is to be made or is being made at each instant. Otherness is seen as the structure and condition of the possibility of worlds, like this otherness of the other that keeps the relation in motion and in dynamics, "like the condition of the perception field, [whereas] the world outside of the reach of present perception has its possibility of existence guaranteed by the virtual presence of an otherness by whom it is perceived; for me, the invisible subsists as real because of its visibility for others" (Viveiros de Castro, "O nativo relativo").

Without otherness, the world crumbles, as it is the principle that constitutes the perception field, but, in turn, in its relationality, it can be modified, making it undertake new and unpredicted configurations. That is, otherness is also the condition of variability of perception matrices that, even though they may be inhabited by the object/subject pair and their representations, they are independent of. It is in this sense that a cosmopolitics of image, like the politics that is exercised in a certain ecology of cinematic practices by incorporating other components, is not interested in the expressive forms of the others, rather in the transformational and transmuting powers they bear. Those potencies of "cosmic events" postpone not just the fall of the sky or the end of the world (Kopenawa and Albert), but also the sadness and misery of the fixation of an image of thought (Deleuze and Guattari). The cosmopolitical covenant and the common cause that emerges from that is not with the other, but with an otherness that it dramatizes, refusing the totalization of a world and, with it, the stagnation of the potency of thinking.

If, in *Abyss,* this dramatization occurred in the encounter with a chromatic vertigo of cosmic molecularities, in *Xapirimuu,* it occurs through the alliance with the Yanomami potencies of thought, which expand the ecology of practices by means of transformational cosmopolitical performativities of productive equivocities (Viveiros de Castro, *Metafísicas canibais*). As we have been insisting, this is never about finding equivalence or even less about promoting homogenizing gestures. As Stengers says, there is the need of a principle of precaution ("A proposição cosmopolítica") to avoid limiting the other and also to, conversely, be able to give way to the otherness that inhabits and traverses it. Not just bringing into existence this possible world emanated by otherness, but, as a cosmopolitical gesture, activating other unpredicted possibles in the plane of experience as processes of mutual transformation and inclusion (Massumi), without ever aspiring to a synthesis, to a unifying project.

Xapirimuu never has the arrogant desire to cover the wealth and diversity of thought and audiovisual potencies presented in *The Falling Sky* (Kopenawa and Albert). On the contrary, in the vertigo, and trying to *make body with* a complex but instigating proposition—to act in spirit—it tries, and that is always a risk, to apprehend a certain transformational potency; in this case, in the audiovisualities, which would involve activating a certain affective atmosphere that, somehow, always uncertainly and in a diffuse manner, would resonate with Yanomami affects. Those are bare threads that may not mean the same that Kopenawa and Albert propose in their book, but, in the cosmopolitical performance that they drive forward and integrate in this ecology of practices, they might be able to affirm the gesture of dreaming the land between the clouds so that the sky does not fall.

The cinema by other means that I believe Yanomami shamans drive forward, Kopenawa in particular, as other modes of cinematic experience, is not only independent from a medium such as the film, but above all it is independent from the assumption of visibility. The matter of this cinema is what the Yanomami call utupë, which, in its turn, composes the xapiripë. This cinematic matter "is both non-iconic and non-visible"; it is a spiritual matter of expression. That is, they are spirits who index "the characteristic affects which they are the image of, without looking like that which they are the image of: they are indices, not icons ... so they would be like the condition of that which they are the image of; they are active images" (Viveiros de Castro, "A floresta de cristal"). And it so happens that a cinema of process can only be interested in "active images," in living materialities—that is, unfinished ones, always making themselves, acting in spirit because they are in constant metamorphism, they are in a state of becoming, they breathe and compose heterogeneous atmospheres. Somehow, this is what is poised to be perceived and felt in *Xapirimuu*, in its diversity of techniques and procedures, without offering an image of Kopenawa's and the Yanomami thought. The "active images" do not unveil or reveal, but, in immanence, they activate, they assemble affective tendencies, unpredicted possibilities of composition. In order to precariously find one's way in this *acting in spirit* we may listen to the propensities of the audiovisual blocks without knowing a priori which direction the index that they carry will throw us at. It is because of this risk and vertigo that, just like *Abyss*, *Xapirimuu* can end and begin at any point.

Since everything can end any instant, then everything comes to life each instant. Each instant—it is our obligation as practitioners of this ecology—we must activate becomings, bring to existence, pour ourselves once and again in the cosmogenetic process of the ecology itself, where everything is active and activating, where everything contains a potency of life, where everything is spiritual force in act and action with the unfolding of the pluriverse in construction and proliferation. Thus, the visualities and sonorities are freed from the constraint of representation. There is no world to be represented, but there is a need to activate genetic forces of the cosmos so that an emergent, always nascent world can be installed.

In *Xapirimuu,* audiovisual composites, like tangles of analog and digital matter-cinema—or so-called animations emerging from the various materialities—feel this need and calling. The audiovisual composites, like all spirits, know that anything that can activate a becoming is welcome. It is not surprising, then, that in *Xapirimuu* lines, wires, glitter, paper, fabric, all are living together joyfully; they are making the cosmopolitical performance endure, the situation-film, in the search for this always fugitive "cosmic event" that gives consistency to the ecology of practices and insists on being the catalyst of the disparate so that the image can avoid being captured. The image, paradoxically, never ceases to capture new rhythms that always make it an other, always yet to be made, always in variation.

As mentioned before, what is of interest here is a cinema that calls for the interval in the midst of the vertigo of visualities and sonorities, even when invisible or inaudible, but which are attractors for the dynamism of thought and perceptive matrices. Certainly, Kopenawa's thought affirmatively contaminates the ecology of practices at hand, where, for instance, a perceptive interval is affirmed here and now in act, independently of the degree of visibility or audibility of the visualities and sonorities of this atmosphere which we are we momentarily traversing in writing and reading.

The Yanomami, and Kopenawa, as practitioners of very particular modes of cinematic experience, teach us that this cinema of process must be open to a constant destabilizing and hesitating action, where, for instance, being worthy of the interval or the otherness that the xapiripë carry, entails, for the shaman, complex initiatory processes where the obligations and responsibilities before the ecology of practices are even more notable and imminent. The literality of always being amid a great cosmic cinematograph in formation is tremendously vertiginous. It is with this vertigo that a cinema of process is called upon to *make body with.* So, by integrating these Yanomami thought forces to the ecology that calls upon us here, it becomes palpable and clearer than ever that we are never dealing with facts or with a nature of things. On the contrary, we are dealing with, and giving ourselves to an emergence of propositions, occasions, and situations that trigger the activation, action, and transformation that render cosmopolitics a politics of affects, always more-than-human, always refusing to serve a lord, or nature (there is no such thing as a unifier), or any transcendence. A politics that, at any rate, would only serve a countertranscendence that Stengers calls "intrusion of Gaia" (Sztutman, "Reativar a feitiçaria") and that we will call "point of view of creation," where, as we insist, ontologies and worlds are made, are composed. A counternature compositionism that only rises because it creates conditions of existence for the diverse, for the being-together of difference, where we are all atmosphere, where we are all medium and are in between.

Considering atmosphere an "aggregate of diverging worlds that do not accept a common representative" (Sztutman, "Um acontecimento cosmopolítico"), we inexorably pour ourselves into a plane of experience inhabited by freely emerging multiplicities where, without transcendence, we gain a higher degree of intimacy with chaos and its always fortuitous ways, enriched in cosmicity. Thus, the obscure precursor of otherness is always chaos; but otherness, as a passage of cosmicity and interval, is the source, the spring which worlds are engendered from. In the interval, as participants and practitioners of this ecology, we have the obligation of caring for this nursery-occasion where sparks of life are in constant eruption, where they are affirmed as the multiplication of active counterimages that make cinema not only a cosmopolitical gesture but also a shamanizing vertigo of transformation between worlds, a bridge toward the outside.

In the encounter, in the creation of partial connections with the Yanomami thought potencies, the modes of occurrence of the plane of cinematic experience gain new experimental shades, making explicit that, like in sorcery and magic, what must be cultivated "is an art of immanent attention, an empirical art that investigates what is good or harmful" (Stengers, "Reativar o animismo") for the ecology of practices and the cosmogenetic process it brings forth. It is in this sense that shamans are radically pragmatic and "experiment with effects and consequences of that which, as they know, is never innocuous and involves care, protection, and experience" (Stengers, "Reativar o animismo") and that is what, somehow, *Xapirimuu* explores by evoking vestiges of the metallic mythology that Kopenawa presents.

As an ethos-ecology, this cinema of process must always become a gesture of care and healing, and therefore of resistance. It is resistance to any will to make matter impotent or inactive. In other words, it is barren, and barred from entering relations and making connections that are different from those already known and expected. Knowing that everything can become matter-cinema, but also that everything can languish; caring and, if necessary, healing are imperative, precisely for that matter-cinema to be materialized—namely, so that it can enter a process of composition, of cocreation, and coevolution. In other words, much like the shaman's body and the shaman's practice of *making body with*, it is about becoming porous, permeable, malleable, open to fractalities and passages of low adherence. That was, perhaps—and always as a risk and vertigo—what guided the situation-film of *Xapirimuu* among indiscernibilities, but also as singularizing modulations of a contingent dramatization, where what was attempted was to hear the secret language of visualities and sonorities, a hearing that does not unveil the secret. If that were the case, we would cease to be the catalysts of the differential and of the unlocatable and unnamable epicenter of the virtualities that sustain and give cause to this cosmopolitics of the image. We listen to the murmur of the cosmos in the abyss that is the interval, but never ask where it comes from or where it is going. It is just about multiplying the surfaces where it can proliferate and modulate. This is at once a simple and complex gesture, as a continuity in the discontinuity, that permanently permeates the cosmopolitical performance.

The Yanomami offer us a very rich concept with the polysemy that the notion of utupë carries and that creates a point of no return in this ecology of practices, by making it enter transformations not just because it makes it embody new components, but because those components, much like the others that already comprise the rhizomatic complex, can no longer stop revolving and shaking the morphogenetic processes. Active images fractally unfold into constant becomings and metamorphisms and they have the force of speculating (Valentim). Counterimages carry fertility: an impregnating force that makes them multiplicity and a flow of heterogeneities. They are alive, they are spirits, they act in spirit, as their agency moves a world so intensively that it does not stop opening into new ways of existence. They are attractors of transformations, pure propensity toward becomings. Utupë makes us remember that we are part of the same lineage, of the same cosmogenetic process, and that, therefore, we are part of the constant intertwining of perspectives that have multifaceted the world, multiplying the inputs and flows of otherness and, with it, of new modes of existence. Unending series of embryonic occasions where, for instance, this occasion-film-paper in act, and what you are reading now, must be utupa sipë or "skin of images" like the Yanomami call the inscriptions on paper.

We insist, with utupë, that a cinema of process must always be willing to be produced through other means, to be multiplied through several and unpredicted surfaces, to be a spirit and spiritual as a result of having too many bodies that overflow their own limits, almost up to the point of bursting and leading to a collapse of the planes of the audible and the visible. A cinema that can reach spectral thresholds, because, as a spiritual conglomerate, "it is only scarce of bodies for having too many of them" (Viveiros de Castro, "A floresta de cristal") that are unstable and, at the same time, disjunctive. A cinema that, in its spectrality, thinks the world; that, in its constitutive metamorphic potential as a cosmic agent, "embodies a structure of thought that is intensely transformational, a-subjective" (Valentim), in which, as Deleuze reminds us, to think is always about creating and differentiating.

The notion of utupë comes forward as a concept, as it is affirmed as a hyperconglomerate of relationships that dynamically condense a quality of vortex and vector for thought. Utupë does not only affect the way audiovisual blocks are composed in *Xapirimuu*, but, since the audiovisualities in this ecology of practices always have a conceptual reverse, in their turn, they nurture the possibility of giving greater consistency to the notions of kino-madology and poetic ethology that populate this plane of experience as critical zones of transversalities between heterogeneities. With utupë, we claim that this kino-madology and poetic ethology are inhabited by thinking beings, by active counterimages, which, as occasions for thought, as occasion-film, paper, concept, like we have been exploring here, detain perspectives through which otherness transpires. And let us not forget that the proliferation of otherness depends on the proliferation and dynamism of perspectives, as spectral interval-operating escapes that sustain the world and the image. The image is not what is to be seen in the world or from the world, but something that sustains it, including our own bodies, for being that outside as a dynamic core and spectral agency, which is a diffractive and nonreflective vibration preventing the stagnation of movement.

In other words, the preoccupation and the obligation of this cosmopolitics is to safeguard the efficacy of the image, which can acquire unpredicted shades that continue to affirm that vertigo is a primordial factor to, in the immanence and through the medium, inhabit the cosmic cinematograph that the world is.

Notes

1. In this text, as an intermedial experimentation, I do not want to subalternize audio-visualities to a discursive regime. I believe in the autonomy and incommensurability between images and words. Hence, the experience of this composition of thoughts demands an oscillation between textual and audiovisual compositions. The reader is invited to watch the short film *Abyss* following this link: https://vimeo.com/54427965.
2. As previously mentioned, in this text, as an intermedial experimentation, we do not want to make audiovisualities subaltern to a discursive regime. The reader is now invited to watch the short film *Xapirimuu* following this link: https://vimeo.com/195130177.

Works Cited

Deleuze, Gilles. *Crítica e clínica*. Editora vol. 34, 1997.
Deleuze, Gilles. *La imagen-tiempo*. Paidos, 2005.
Deleuze, Gilles. *Lógica do Sentido*. Perspectiva, 2015.
Deleuze, Gilles, and Félix Guattari. *O que é a filosofia?* Editora vol. 34, 1992.
James, William. *Essays in Radical Empiricism*. Dover, 2003.
Kopenawa, Davi, and Bruce Albert. *A Queda do Céu*. Companhia das Letras, 2015.
Massumi, Brian. *O que os animais nos ensinam sobre política*. n-1 edições, 2017.
Orlandi, Luiz B. L. "Revendo nuvens." *ClimaCom*, vol. 7, 2016, pp. 91–117.
Savransky, Martin. "Pensar el Pluriverso: Elementos para una filosofía empírica." *Diferencias*, vol. 1, no. 8, julio de 2019, http://www.revista.diferencias.com.ar/index.php/diferencias/article/view/180.
Spinoza, Baruch. *Ética*. Relógio D'água, 1992.
Stengers, Isabelle. *In Catastrophic Times: Resisting the Coming Barbarism*. Translated by Andrew Goffey, Open Humanities, 2015.
Stengers, Isabelle. "Introductory Notes on an Ecology of Practices." *Cultural Studies Review*, vol. 11, no. 1, 2005, pp. 183–96.
Stengers, Isabelle. "A proposição cosmopolítica." *Revista do Instituto de Estudos Brasileiros*, no. 69, 2018, pp. 442–64.
Stengers, Isabelle. "Reativar o animismo." *Cadernos de Leitura*, vol. 62, 2017, pp. 2–15.
Strathern, Marilyn. *Partial Connections*. Rev. ed., Altamira, 2005.
Sztutman, Renato. "Reativar a feitiçaria e outras receitas de resistência—pensando com Isabelle Stengers." *Revista do Instituto de Estudos Brasileiros*, no. 69, 2018, pp. 338–60.
Sztutman, Renato. "Um acontecimento cosmopolítico: O manifesto de Kopenawa e a proposta de Stengers." *Mundo Amazónico*, vol. 10, no. 1, 2019, pp. 83–105.
Valentim, Marco Antonio. "Utupë: A imaginação conceitual de Davi Kopenawa." *Viso Cadernos de estética aplicada*, no. 24, 2019, pp. 193–216.

Viveiros de Castro, Eduardo. "A floresta de cristal: notas sobre a ontologia dos espíritos amazônicos." *Cadernos de Campo,* vol. 15, no. 14–15, 2006, pp. 319–38.
Viveiros de Castro, Eduardo. *Metafísicas canibais: Elementos para uma antropologia pós-estrutural.* Ubu Editora, 2018.
Viveiros de Castro, Eduardo. "O nativo relativo." *Mana,* vol. 8, no. 1, 2002, pp. 113–48.
Whitehead, Alfred North. *The Concept of Nature.* Cosimo Classics, 2007.
Whitehead, Alfred North. *Process and Reality* Free Press, 1978.

12

Brazilian Afrofuturism, Climate Apocalypse, and Heuristic Function

Patrick Brock

Science fiction (SF) dystopias in which the rich can enjoy the spoils of planetary catastrophe after bunkering down are hallmarks of the genre. But what about stories where the postapocalypse consists of the hard work of rebuilding? Such is the case of the two-part novel *Brasil 2408* by Brazilian Afrofuturist author and educator Lu Ain-Zaila (then–pen name of Luciene Ernesto), which takes the Afrofuturist promise of a counterfuture and upheaval of the modern, white temporality and its assumptions (Neyrat 121) into a highly political and economically utopian space.

Adam Trexler argues that "climate novels have a role to play in our collective accounting of the Anthropocene" (237). Meanwhile, as often pointed out about the Afrofuturist movement (Eshun 288), the experience of Black people is already postapocalyptic. Moreover, Afrofuturist writers elsewhere are taking upon themselves the task of creating environmental counternarratives (Winter 189–90) to the "There is no alternative" argument for neoliberalism and eternal growth, recently rebranded as "sustainability." In this sense, studying the work of Ain-Zaila, which emerges from the periphery of Brazil's everyday dystopia, can offer a window into a different imagination of climate change and its redemptive potential. This chapter will argue that the two-part novel[1]—*(In)Verdades: Uma heroína negra mudará tudo* ([Un] Truths: A black female hero will change everything) from 2016 and *(R)Evolução: Eu e a verdade somos o ponto final* ([R]Evolution: I and the truth are the end point) from 2017—while not offering a definitive solution, presents an educational how-to that is both problem-solving action and method.[2] Ain-Zaila uses a multifaceted patchwork of fictional news reports, textbooks, first-person points of view, SF tropes, and aspects of political thriller and police procedurals to imagine a refoundation of Brazil and pose a response to climate change. Ultimately, transformation is dependent on a collective and transparent effort of nation-building, as opposed to a market-based approach focused on supposedly liberating individualities.

The rise of a Black middle class and affirmative action policies in the early 2000s (Duarte et al. 98–99) have laid the conditions for today's Afrofuturist literature boom in Brazil. Part of a broader shift in the production of SF and utopian thinking in the country, the movement's growth also stems from the wide availability of new communication and information technologies, but it is still firmly grounded in a long-standing tradition of grassroots activism (Brock 321–22). Following in the footsteps of earlier activists who mobilized against the Brazilian myth of racial democracy (Jones 16–17, Melo 264), today's Brazilian Afrofuturists intervene in the speculative tool kit of SF and its aesthetic to claim a space that has been long denied. Moreover, Afrofuturism's vigorous hold on Brazilian culture shows how it is a politically conscious aesthetic activism, which originated in the North American Black diaspora and transcends borders and cultures as a form of transnational "networked consciousness" (Lavender III 7).

Joel Burges and Amy J. Elias question whether "temporality itself might be something that always eludes complete co-optation by capital, something on a different categorical or ontological level leading to multiple fractures and sites of resistance" (12). Afrofuturism's temporal problematization fits that notion, as Amelia Groom points out that the moment diverges from the earlier "postmodern historicity" in that "rather than a winking postmodern pastiche of appropriated styles," these new experiments mark a "thickening of the present to acknowledge its multiple, interwoven temporalities" (Burges and Elias 19–20). Instead of accepting White temporal narratives where time moves in a straight line to progress, Afrofuturists establish a chronopolitics that embraces the complexity of history and presents it as a multilayered and ongoing process.

Fixing the Monopoly on Possible Futures

> There is a popular saying: One day nature will grow tired and will find a way of seeking equilibrium.
> —Lu Ain-Zaila, *(In)Verdades* (23)

Ecologically themed novels have a long history evolving in tandem with our relationship with nature—from visions of capitalistic exploitation to the current concerns with planetary degradation (Asselin 452). Climate fiction has expanded recently as factions of society continue fighting to overturn the established discourse that profits matter more than planetary destruction (Trexler 3). SF itself also has a broad corpus of works dealing with climate change, as well as Latin America's contribution to the genre. López-Pellisa and Kurlat Ares (476) cite several environmentally conscious novels published as far back as the early twentieth century, including such ice-age narratives as Eduardo Herrera's *El fin de la raza* (Peru, 1910) and *La Tierra Dormida* (Chile, 1969) by Ilda Cádiz. More recently there is the work of Claudia Aboaf in *El Rey del Agua* (2016) and *El Ojo y La Flor* (2019), both set in a near-future Paraná Delta, just north of Buenos Aires, where extractivism has turned to water as a resource more precious than oil after climate change. Equally invested in the imagination of the postapocalypse is Michel Nieva and his *La Infancia del Mundo* (2023), where the final melting of polar ice by 2197 turns the Pampas into a new Caribbean.

Given the complexity of our current crisis and how entangled with economic structures it is, world-building and world-making arguably take on an urgency for truly radical economic innovation that progressive SF can arguably envision (Shaviro, "Unpredicting" 8; Pratt 9). And the intersection of narrative and world-building (and making) in our political economy and the environment is one element that the novels of Ain-Zaila explore in innovative ways.

Capitalism has long relied on its own narratives: the meritocracy of self-made men, the free hand of the market, and trickle-down economics are just some of the arguments employed to justify its ways of organizing society and rewarding (or not) different stakeholders. In a timely anthology on SF and economics he edited in 2018, William Davies questions economic mythmaking with an observation about a foundational claim of Austrian economist Ludwig von Mises in 1920. Mises argues that in an industrial economy lacking a system of monetary prices "there would be only groping in the dark" (Davies 1). Davies counters this by asking whether we are truly able to imagine a working alternative to money and the organization imposed by market forces. During Mises's time, the challenge facing economic liberals like him was how to sell the notion of free markets as modern following the demise of laissez-faire classical liberalism in the nineteenth century. As David Harvey shows in *A Brief History of Neoliberalism* (2005), this quest evolved into the argument that strong property rights, free markets, and free trade were liberating ideals for individuals. Today these propositions are increasingly used to justify the dystopian lives experienced by a growing number of people, and the utopian lives of the few. To counter that prevailing paradigm, the challenge now is to mount a renewed effort of imagination that rejects the idea that oppositional proposals cannot exist. Mark Fisher argues that understanding how the underlying assumptions of our economy themselves are constructed fictions should inspire efforts to consider other possibilities: "Capital's economic science fictions cannot simply be opposed; they need to be countered by economic science fictions that can exert pressure on capital's current monopolisation of possible realities. The development of economic science fictions would constitute a form of indirect action without which hegemonic struggle cannot hope to be successful" (xiii).

Therefore, the conditions of our highly financialized economy, in which "contemporary money is so abstract, so divorced from physical matter, that its ability to constrain us can seem bizarre" can become "ripe territory for the inventive imagination to play around with" (Davies 8–9). It is here that Ain-Zaila offers a window of imagination. In her fiction there are no cannibalistic road warriors, vestigial civilizations, or starving masses dwelling in radioactive deserts, nor partially submerged but still gleaming cities to continue the work of globalized financial capitalism. There is just the everyday struggle for survival that marginalized peoples of the Global South have experienced for all their lives. Life goes on despite the apocalypse, which has already happened anyway—diegetically and in the broader context of the Black, Latinx, and Indigenous experience with colonialism and systemic racism (Sanchez-Taylor 2021); people still fall in love, buy groceries, and hope for a better future. This time, the "tragedy of the commons" (Nijhuis) is what has been canceled. Unlike Amitav Ghosh's observation about our current politics as merely a performative act which finds its ability to influence power increasingly subdued (129), the characters of Ain-Zaila's duology will not stand down. Imbued by a sense of agency, they fight both inside and outside the body politic. Her speculation about the future, like other efforts (Milner and Burgmann 191), addresses the present of climate change through calls for collective mobilization.

Speaking to Ain-Zaila in 2021, she said the initial idea was to depict the aftermath of a nuclear apocalypse. Inspired by the environmental realities of her hometown of Nova Iguaçu, in the Baixada Fluminense area of Rio de Janeiro, which is prone to flooding and extreme heat, and reports of tornadoes appearing in the south of Brazil (to which Ghosh also alludes in *The Great Derangement*), she shifted to climate change. The impulse to write the book also came from a visit to Rio's Book Biennial, where she could not find books written by Black women. Guided by her experience with Black activists while studying for the university entrance exam in the early 2000s at community schools for low-income students, and by the literature groups that had been growing in Baixada at the time, she took up the task of writing the novels in her spare time while working as a pedagogical coordinator for a school in Rio. Ain-Zaila also created the cover, design, and illustrations of the books, drawing from her experience of working with a printing business and knowledge of computers from her vocational education in high school, and went on to self-publish them in print for sale at events and through Amazon, where an ebook version became the most significant source of sales during the pandemic.

Ain-Zaila reported she had little access to Afrofuturist or SF literature while growing up. Comics, manga, and TV broadcasts were the most accessible cultural products. But African culture and the example of Black Brazilian activists were always present from her experiences in Baixada. As a social activist, she joined the movement to push Congress to pass the country's first affirmative action law in 2003. Once the work of authors like Octavia Butler became available in the second half of the 2010s, she says the effect was eye-opening. After *Brasil 2408* she published more fiction as well as essays and papers on Afrofuturism, becoming one of the movement's leading voices in Brazil, with her work and research increasingly involving the educational instrumentalization of the movement (Ain-Zaila, "Prefácio").[3]

Postapocalyptic Nation-Building

For Ena Dias da Silva, the protagonist of the novel, the future seems bright. In the first book of the duology, the recent college graduate is heading home to await the results of her final examination, which will determine whether she can join the elite Brazilian District Forces, a combination of law enforcement and scientific task force assigned to maintain order and survey the now inhospitable wilderness beyond the cities. She wants to join her parents in government service, particularly her late father, who was also part of the Forces. Readers are treated to the seamless banality of Brazilian interstate travel in the year 2407: secure but not overly intrusive onboarding, a quick flight, and then a train ride home, where two pets and a note from her mother await her, warning that the fridge is empty.

But we soon realize there is a demanding universe of economic arrangements and climate necessities underpinning her existence, on top of a tragic past. Consumption is tightly controlled; wealth accumulation is nearly outlawed; the government is preparing to centralize control of an economy pegged to scarce biological, natural, and water resources. Everywhere there is a pervasive sense of economic downsizing and devolution due to climate change. Earth's immediate orbit is, for instance, littered with too much space trash, making it economically impractical to launch more powerful satellites. It is also an austere Brazil in which the consumption of alcoholic beverages in public streets is outlawed and capital punishment made legal, denoting the novel's intense preoccupation with the rule of law. Also, dark forces seem to be disturbing the fragile balance of scarcity and rebuilt governance as criminals keep hacking the system with relative impunity. On top of it all, Ena is still struggling to accept the tragic death of her father, who was killed a decade before in a terrorist attack blamed on the "nonidentified" dissidents (*[In]Verdades* 89).

Before the plot's introduction, Ain-Zaila uses a patchwork of different texts to build her world: following a widespread environmental catastrophe in 2198 that left unclaimed bodies piled up, Brazil has reassembled itself into three megaregions where people live clustered in cities. With the destruction of the Brazilian coast, the country's most populous area, and consequent demolition of its buildings to create a sea wall, climate change led to two hundred years of chaos and strife. Efforts in the twenty-second century to reform energy generation prove fruitless against destructive climate change. Floods, imbalanced ocean currents, distortions in air humidity, price volatility, water scarcity, and torrential rainfall generate nonstop chaos. There is no way to bury the dead or search for survivors, with cremation being the only remedy by 2270. Brazil loses about 40 percent of its landmass due to contamination, flooding, and tornadoes rendering the south region uninhabitable. Powerful storms that can form on short notice force people to take refuge in the rest of the country. Venturing out into the wilderness is a job for highly trained forces or those on the margins of the law, turning aggressive ecological surveying and conservation into a big concern. Outside cities, the wilderness is either too dry, polluted, or dangerous for people.

Around 2268, efforts to begin mapping the transformed territory and establish new borders embody a renewal of the nation-state upon different bases. The country had to reorganize itself into an unstable equilibrium where some rebel against the technological totalitarianism of a surveillance society and its mandatory personal devices. Also, while stable, life for those who accept this tightly regimented surveillance is still threatened by criminality, high-level corruption, and unstable weather. Each of the three districts has three representatives with autonomy, who gather one week every month to discuss and vote on national matters. They are elected from the population for three-year terms and are entitled to no special privileges beyond state-sponsored security. The goal is to have ordinary people serving, and they can be recalled at any point by vote.

The ethics of resource scarcity and of technology are critical concerns. By law, no water or energy resource can be fully privatized, and all of it must be sustainable. Technology must be used to ensure quality of life, and all forms of housing must tend to collective needs. The distribution of food is also determined by law, with a public exchange system being set up to manage resources through the Public Distribution Control Center. At the same time, purchases go through the Integrated Communication Device (ACI), a smartphone combining media reader, wallet, and identification app. Each purchase is registered at the person's ACI and connected to the rationing system, which the population can monitor in several public panels spread around the district. But people still find ways to cheat the system, forcing the authorities to compensate vendors for unregistered sales that generate losses. Ena's mother, Naira, creates and oversees this system, in which currency reverts to the social technology role that had been replaced by the notion of money as a "thing" that "we treat . . . as a mathematical truth rather than a social choice with often disastrous consequences" (Vint 63). Within this new framework of money pegged to natural resources, the sense of shared wealth becomes the nexus of cooperation for this rebuilt society in a world of scarcity. But the estrangement caused by the total surveillance required to manage this economic system affects even Naira. "We already have so many things being controlled that I can't get used to the idea of someone picking even my carrots" (*[In]Verdades* 60), she says while explaining her need to go to the market.

In this future society surviving amid scarcity, equality is radically imposed: "If we only have a few beans, then all of us will have fewer beans, and anyone who doesn't respect this equality should be seen as an enemy, because if they want more, someone else will not have food, and then our civilization is put into question" (*[In]Verdades* 106).[4] Or, as defined in another section: "After the Great Climate Change, there was no more middle, high or low class but survivors in an adverse environment" (39). The number of children a person can have also is limited. A sense of emergency pervades this new reality, with in-person classes only being allowed for children who are old enough to be trained to deal with emergencies. The plot depicts a potential future where postapocalypse has led to a national refoundation instead of something more akin to the *Mad Max* series of films, for instance; but it also shows that some are more equal than others, as some form of economic elite survived the apocalypse and is now scheming to protect its position, showing how the wealthy often are able to insulate themselves from crises and recover better (see Hallegatte and Walsh).

Energy and renewability concerns permeate the technologies described in the novel. All buildings have mandatory solar panels for distributed generation, while their height is limited to eight stories due to the unstable weather. Mobile solar plants also make up the new infrastructure. Packaging for the cadets' rations is made from biodegradable material, which dissolves in a matter of days once discarded. "Climatizers" located in every block help regulate the temperature and humidity. They have become legacy technology at the story's start, operating for decades but still necessary to survival ([In]Verdades 37). Ena considers whether they will one day become museum pieces once the climate recovers its balance. The observation shows a temporal paradox: technologically speaking, none of these are particularly futuristic. Radiofrequency means of payment or facial recognition cameras, computer networks surveying data in real time, distributed generation, wind farms, and solar panels are all commonplace. This is where the novel's radical futurity resides: showing a future Brazil that might as well be happening now if it were not for hegemonic interests maintaining the use of carbon-intensive energy sources and the current system of market-set prices and distribution.

Trauma and Collective Action

Afrofuturism initially takes a backseat to the cli-fi, melodramatic, and thriller elements of the novel's hybrid form. Yet it remains an essential backdrop in the ways it portrays the agency of Black people and their positions in power as racism becomes taboo and is represented in museums as part of the problems of the past that led to the climate destruction. The centrality of trauma (the collective one of the climate disaster and the protagonist's personal loss) and the challenges that Ena faces (resistance from superiors, sabotage, and veiled racism), as well as the position of the undocumented faction, are central Afrofuturist elements. Epistemological diversity efforts are another strategy, with the novels permeated by African cultural elements like the Adinkra symbols (Adom et al.) as well as historical facts and flora from the continent.[5] Ena's hair is described as having a "uniquely random order which any attempt at organization, as small as it can be, would destroy its harmony" ([In]Verdades 31)[6].

The duology carries at least two of the five elements identified by Duarte et al. as recognizable traits of Afro-Brazilian literature: authorship (life experience marked by obstacles and the assimilation of trauma) and the focus on a specific reading public (111–19), especially the latter, since the accessibility afforded by online media creates opportunities for the digital inclusion of marginalized peoples. But the work of Ain-Zaila goes against Duarte et al.'s notion that Afro-Brazilian literature is an outsider and does not locate itself "within the romantic ideal of instituting a national spirit" (119). Ain-Zaila meditated on the issue, writing an essay in 2018 where she recognized the racist and colonialist roots of SF and how "our backward national project remains in full force" (Ain-Zaila, "Ficção" 60). By imagining a new kind of nation-building, the duology adds another dimension to the already transformative spirit of Afrofuturism, with its struggle for agency in affirming identity and occupying spaces previously denied. By alluding to geographic and ethnic concerns, the novel counters the notion that it stands outside the country's literary tradition, as had been posited about both Afro-Brazilian literature (Jones 16) and SF in the country in general. Moreover, as argued by Kurlat Ares and Rosso ("Historia" 10–12; "La ciencia ficción" x), Latin American SF has always sought acknowledgment in nation-building efforts as much as its more prestigious literary modes, offering both a social laboratory and an alternative for a formative system of knowledge and identity often captured by intellectual elites in what Ángel Rama has famously called the "lettered city" (Colombi 83–91).

One of the most critical signs of the novel's struggle for agency and visibility is Ena's paramilitary training. The armed forces have always played a crucial role in Brazilian politics (if not in the whole continent), pushing for a republic in the late nineteenth century, modernizing the state during the Vargas era (even as it descended into authoritarianism), and suffocating democracy and committing crimes against humanity as part of the proxy struggles of the Cold War. The role played by law enforcement is different here, as more of a humanitarian and scientific force, and Ena's presence in their midst represents the desire for a role in the efforts of security, law, and order, areas in which Black people are traditionally denied agency in Brazil. Again, it embodies a desire to take a central role in the narrative and nation-building efforts. Thus, government is a core element of the novel, either in the form of its middle managers or officers, or the inner workings of its high council. Reconstruction is led by a highly streamlined government akin to the Federal Council of Switzerland (which has seven members acting as a collective head of state) and combining roles of the executive and legislative branches.

Later, it turns out that the only way to defeat a plutocratic conspiracy lurking in the revamped body politic ultimately resides in radical economic transparency and collective action: "The earlier damage is a process that only the collective, in a continuous effort over many centuries, will be able to mitigate" (*[In]Verdades* 75). The argument for collective action repeats itself in the novels: "Ideas or speeches don't overthrow governments or democracies when stated by a single voice, but if enough people are hearing, agreeing and sustaining them, everything can change" (65), or "for many centuries we, Brazilians, didn't participate in the process and were swallowed by the results, but it seems like we finally got our acts together and are attentive and involved" (191). The focus on the collective is a welcome respite from today's hegemonic individualism since, as Fisher reminds us, neoliberalism's "rhetoric of releasing individual potential obfuscates its suppression and fear of collective agency" (xxi). The social and economic organization of this future Brazil noticeably evokes earlier socialist experiments in the Soviet Union and Chile that sought to use computer networks to manage the economy, replacing market forces.[7] In Chile, once the half-finished effort was discovered by members of the military dictatorship (1973–90) of Augusto Pinochet, it was quickly replaced by another form of economic organization emerging at the time: neoliberalism (Tironi and Barandiarán 305; Medina 211).

Utopia and the Heuristic Imagination

The first part of the novel focuses on Ena's training for the Brazilian District Forces and her coming of age, but, as she graduates and joins the Forces and the plot evolves in the second part, she begins uncovering a conspiracy involving high-ranking members of government, law enforcement, and the remaining business interests of postapocalyptic Brazil. While she investigates and builds allies, the search for her personal truth blends with the broader quest for transparency and agency. When it becomes evident that the conspirators have insider knowledge, she abandons the District Forces and takes refuge with the nonidentified. Despite living in the margins of society, they still have representation in the political body, albeit a fragile one. Their civil disobedience against what they consider excessive intrusion into their privacy, while businesses and the government are spared transparency, forms one side of the backbone of the novel's social conflict. More than economically marginalized, they represent those who rebel against Brazil's inability to change and develop a more equal society. Even Ena must remind herself to erase their otherness: "I was so used to living in the system that I completely forgot the challenges they face for not joining it" (*[In]Verdades* 186). On the other side of this social conflict are the heads of businesses that undertake major construction projects, generate electricity, and control large industries, forming a monopolistic conglomerate. Along with some of the founders of the new Brazil, the entrenched neocolonialist oligarchy includes the remaining elite from the before-times.

New forms of money tend to destroy the ways of governance based on previous economic systems (Vint 61). Therefore, once the new economic management system created by Ena's mother begins operating, the monopolistic elite and their allies in government set out to sabotage it. In the end, they are the same Brazilian oligarchy intent on protecting its position through national stasis: "There is an immense filth under our splendid cradle" (*[Re]Evolução* 70), states one character in a reference to the Brazilian national anthem. The reactionary forces of this future are also symbolized by Dantas, one of two representatives from the Northeast district. "You people, the non-documented, oppose the progress of humankind!" he utters against their representative in the governing council, who responds: "A stupid [type of] progress brought us here." Despite all that has happened, the narrative of never-ending progress still survives, a critique that is recurrent in the novel, which complains of "a world corroded in a pitiful six centuries of human modernization" (*[Re]Evolução* 50). As the governing council prepares to decide whether to allow the nondocumented to have official votes and proportional representation, the oligarchical groups scheme and sabotage to maintain their privileges in the current order. Once Ena finally uncovers evidence of the conspiracy, revealing it in a viral video, she sets the stage for the next phase in the struggle for change. The masses advance on the main government building while the conspirators retreat to the Annex of Honors, a monument to those killed in the line of duty, where they had hidden the secret servers being used to sabotage the economy and divert funds.

The work of Raymond Williams helps classify the novel's operating mode. Afrofuturism has been equated (Burrows and O'Sullivan 220–21) with one trope he considered in 1988 (Milner, "Utopia" 202): that of space anthropology, a variation of travelers' tales in which new tribes and new patterns of existence are considered. In this sense, novels like Ain-Zaila's open new possibilities because, as Williams puts it, they are "finding in new ways to shape of an alternative, a future, that can be genuinely imagined and hopefully lived" through an "education of desire" (Milner, *Tenses* 102, 124). Chattopadhyay's splitting of the field according to a typology of SF, future fiction (FF), and architectural fiction (AF) helps understand what he defined as heuristic about these and other CoFuturist narratives, particularly how they can reopen possibilities through the prototypes they produce, which "may or may not be immediately realisable but their presentation and speculative design are a prerequisite for future development" ("Speculative Futures" 298). Therefore, these imagined futures are heuristic because, as Chattopadhyay puts it, they create conditions "through which we may renegotiate the possibilities, the potentials, the demands, the horrors, and the pleasures of our various presents" ("Fictioning the Futures" 7). Indeed, recent scholarship has sought to distinguish the type of imagination at stake in more direct terms, calling it a critical utopia/dystopia that replaces perfectionism with self-reflection and plays with ambiguity (Seyfert 2). Either through the ideas of Williams or the more contemporary propositions from Moylan and Sargent (Milner, "Utopia" 209), the duology can be said to carry this ambiguity of a utopia which "must be fought for" (Milner, *Tenses* 103). Moreover, like other science fictional texts by women over the last thirty years, the duology "allows readers and protagonists to hope" and "opens a space of contestation and opposition" to women and other groups pushed out of the center of hegemonic discourse (Baccolini 520).

As Susan Watkins (1–40) points out, women's postapocalyptic fictions from the UK and North America since the turn of the millennium have diverged from the subgenre's tradition of depicting male survivors' longing for the past or trying to rebuild it, by turning to imagining different futures after the destruction. *Brasil 2408* can be considered a distinctive Latin American contribution to that trend, sharing their emphasis on the maternal imaginary and women's relations to each other, for instance, or blaming technoscience, human carelessness, and capitalist exploitation of natural resources for the apocalypse, while inverting the relationship between nature and the nation-state from one of subjugation by the latter; in *Brasil 2408*, it is humanity who must adapt to nature by inventing a new society after the apocalypse. The author chose not to explore the agency of the environment in the duology, and nature ends up taking a passive role as a toxic backdrop to the political and economic world-building. Nevertheless, by emphasizing a socialized economy pegged to natural resources, Ain-Zaila still highlights the importance of centering nature in the construction of possible futures.

Science Fiction as Shared Intelligence

Some of the proposals raised in the novel (different types of price-setting mechanisms, limitation of wealth accumulation, and number of children per parent, among others) align with suggestions brought forward in the late 1990s to address large-scale problems stemming from the unsustainability of Western civilization that were already becoming evident at the time, including potential scarcity. Instant monitoring of real-life data is even being touted more recently as a potentially helpful pathway for economic management. The resurfacing of cybernetically enhanced economic management systems to replace free-market frameworks, along with the simplification of the legislative and executive branches into a council-like body more easily swayed by its voters, are two sides of these potential solutions imagined by Ain-Zaila representing the heuristic function at play. While the former breaks with the dynamic of Western capitalism by putting nature rather than financial capital at the center of the economy, the latter deals with the ossified political structures of Brazil's coalition presidentialism, whose venality and self-interest stand in the way of finding collective solutions. By complexifying the solutions to the scarcities generated by the climate disaster, Ain-Zaila skips the naive assumptions that if humankind would only return to a kind of tribal living, all civilizational problems would be resolved; nor does she ignore the complexities of unsustainable transportation or agricultural and production infrastructures (Trexler 237).

Jasanoff and Kim identify the technoscientific imagination as a social practice, taking the widely accepted role of imaginaries in social theory as the starting point to coin the term "sociotechnical imaginaries" ("Containing the Atom" 120). Jasanoff also questions the role that science and technology can play in converging the "individual subjective self-understanding to a shared social and moral order" and notes that acts of resistance can also draw from the shared tools of "technoscientific imagination and instrumental action" ("Future Imperfect" 5–10). I argue that Afrofuturism does that by instrumentalizing the affect-inspiring technosublime (Csicsery-Ronay 160) of SF and the affordances of platform capitalism:[8] easy communication and dissemination atomizing the publishing industry and letting everyone have a voice.[9] As both a movement and an aesthetic, Afrofuturism takes agency over speculative thought and the accompanying tools and tropes of SF (e.g., the cognitive games, technoscience, utopia, dystopia) to push for a more complex understanding of time and history. However, any truly radical social construction of meaning must also occur at a level beyond what is afforded by internet communication tools in order to escape control in meaningful ways—hence the importance of community practices in the real world, such as those of Ain-Zaila: being activist in her neighborhood and then writing and publishing the Afro-Brazilian literature she found wanting.

This interpretation accepts that SF already constitutes a set of practices embedded in several communities. Steven Shaviro has proposed the neologism "discognition" to explain what he sees as a mode of sentience in SF that disrupts and exceeds, but also subtends cognition. When Shaviro used process philosophy to understand how sentience is imagined in SF as a way of challenging our notions of what is sentient, he also was looking for novel ways of thinking and understanding the nonembodied life of concepts. "My working assumption is that fictions and fabulations are basic modes of sentience; and that cognition *per se* is derived from them and cannot exist without them" ("Discognition" 8–11).

As a concept that has been present in the imagination of the world since the 1950s with the Soviet and Chilean projects and resumed in the vision of a Brazilian Afrofuturist author, these ideas of cybernetic economic governance carry elements that can resemble at least some form of sentience in the way that shared ideas in a culture surpass the individuality of those problematizing them. Ideas gain sentience by reappearing in fiction and art after they were proposed and apparently defeated in our temporality and are envisioned and reenvisioned by readers who become writers and beget more writers, and so on. These notions represent a discognition no less than the notion of a collective organization toward the achievement of political goals or cybernetic economic governance. These ideas are indeed alive and in dialogue with readers, writers, and artists like Ain-Zaila who problematize, imagine, and put them into practice in their fictional worlds.[10] They exist outside us even as we borrow them; they have a life beyond our existence and cannot be controlled by a single individual.

Will all of this make any difference when it comes to effecting actual change? The nascent field of empirical ecocriticism tries to answer that by employing the methods of sociology and psychology; it finds, for instance, that although cli-fi reading may have statistically nonsignificant effects in the long term it can improve the sense of agency and hope of readers (Schneider-Mayerson et al. 9, 21, 23). Capitalism has proven adept at bringing into its fold politically transformative aesthetic works, and the game *Cyberpunk 2077* is an excellent example, but not the only one: think of all the entertaining dystopias available on Netflix, disturbing enough to catch one's attention but still tame enough to avoid triggering deeper introspection. The difference arguably comes from synchronizing the gap between individual action and collective agency, as shown by the power of community-based groups using theater in Upstate New York to raise awareness about the impacts of climate change through participatory artistic outcomes (23–24). In 2018 Shelley Streeby showed that there was a growing connection between SF and activism (119–20). Whether through Octavia Butler's memory work on climate change and other efforts to SF world-building to challenge their erasure from the imagination of the future, they all seem to converge community and political intent with hope and speculative thinking (Streeby 24, 30). Further research could uncover more evidence of how the heuristic function, politicized art, and contemporary futurist movements can work for social change by empowering large numbers of people to turn individual expression into meaningfully collective action.

Notes

1. This result is part of a project that has received funding from the European Research Council (ERC) under the European Union's Horizon 2020 research and innovation program (grant agreement no. 852190).
2. These and all other translations of the novel are by Patrick Brock.
3. Her bibliography can be found here: http://lattes.cnpq.br/8953283041416583.
4. This scenario presupposes the extended aftermath of a chaotic period following highly destructive climate change, where a degrowth philosophy is imposed as the Brazil of 2408 struggles to find a new political economy. Research has found a diversity of responses and outcomes to disasters even though "myths" surrounding those still influence assumptions both of positive and negative behavior (Aguirre). Therefore, new metrics are required to differentiate resilience and outcomes amid rich and poor and tailor disaster response policies accordingly (Hallegatte and Walsh 127–28).
5. The sankofa, meaning it is never too late to go back and pick what was left behind or take the past into account, features prominently not only in these novels but throughout her work.
6. The empowerment of Afro hair is essential to Brazilian Black activism, standing in opposition to the whitewashing pressures of Brazilian culture (Brock).
7. While the Soviet effort became bogged down by internal competition among the bureaucracy, the government of Chilean president Salvador Allende (1908–73) had an even more ambitious computer-based project to manage the economy and prove that socialism could adequately supply its citizens. It even featured dials at every voter's home to express satisfaction or dissatisfaction with the nation-state. Called Project Cybersyn and envisioned by British cybernetics expert Stafford Beer, it would operate a telex machine network at government enterprises relaying real-time data on production and input availability. The data would be analyzed by a simulation software dubbed CHECO (Chilean Economic Simulator) using Bayesian inference to forecast the outcome of decisions. All of this would be supervised

from an operations room in Santiago where government managers could see all data and prepare responses to potential economic emergencies, all in line with the government's expectations of mass resistance from the business sector to its socialization plans. However, the room's sophisticated design was betrayed by a lack of functionality, since its monitors only showed static slides, betraying its intersectional nature between Science Fiction and reality. Also, while the United States had around forty-eight thousand general-purpose computers installed by 1970, Chile only had fifty spread out between government and business, and most were outdated. For a deeper look at Cybersyn, see Medina 2014, while a visual description of Cybersyn is available here: https://99percentinvisible.org/episode/project-cybersyn/.

8 Csicsery-Ronay defines the technosublime as a sense of wonder arising from technical things and knowledges, which operates in tandem with the grotesque to generate an embodied response. He borrows the term from the work of David Nye (American Technological Sublime).

9 The way in which the mostly US-based tech companies operate by combining software and hardware to create a platform where communications, transactions, and services can occur—but also information can be extracted from individuals and sold to advertisers or other parties, extracting value from apparently free activities performed online. See Srnicek (Platform Capitalism).

10 The video art piece converges the same elements of Science Fiction, political economy, and cybernetic governance using narration, subtitles, stock footage, and LIDAR technology. See Bahar Noorizadeh, "After Scarcity," 2018, accessed May 10, 2021, https://vimeo.com/296563987.

Works Cited

Adom, Dickson, Eric Appau Asante, and Steve Kquofi. "Adinkra: An Epitome of Asante Philosophy and History." *Research on Humanities and Social Sciences,* vol. 6, no. 14, 2016, pp. 42–53.

Aguirre, Benigno E. "The Myth of Disaster Myths." *Oxford Research Encyclopedia of Politics,* June 30, 2020. Oxford UP. Accessed 29 Mar. 2024, https://oxfordre.com/politics/view/10.1093/acrefore/9780190228637.001.0001/acrefore-9780190228637-e-1616.

Ain-Zaila, Lu. "Ficção Científica no Brasil: Um Caso de Estudo do Projeto de Nação." *Fantástika 451,* vol.1, 2018, pp. 55–61.

Ain-Zaila, Lu. *(In)Verdades: Ela Está Predestinada a Mudar Tudo. Duologia Brasil 2408,* vol. 1, Kindle ed., 2016.

Ain-Zaila, Lu. "Prefácio." *A Formação do pedagogo mediada por tecnologias educacionais afrofuturistas,* edited by Helena do Socorro Campos da Rocha. IFBA, 2020.

Ain-Zaila, Lu. *(R)Evolução: Eu e a Verdade Somos o Ponto Final. Duologia Brasil 2408,* vol. 2, Kindle ed., 2017.

Asselin, Steve. "A Climate of Competition: Climate Change as Political Economy in Speculative Fiction, 1889–1915." *Science Fiction Studies,* vol. 45, no. 3, 2018, pp. 440–53.

Baccolini, Raffaella. "The Persistence of Hope in Dystopian Science Fiction." *PMLA* 119, no. 3, 2004, pp. 518–21.

Brock, Patrick. "Brazilian Afrofuturism as a Social Technology." *The Routledge Handbook of CoFuturisms,* edited by Bodhisattva Chattopadhyay, Grace Dillon, Isiah Lavender III and Taryne Jade Taylor. Routledge, 2024, pp 319–29.

Burges, Joel, and Amy Elias, eds. *Time: A Vocabulary of the Present.* New York UP, 2016.

Burrows, David, and Simon O'Sullivan. *Fictioning: The Myth-Functions of Contemporary Art and Philosophy.* Edinburgh UP, 2018.

Chattopadhyay, Bodhisattva. "Fictioning the Futures of Climate Change." *Fafnir -Nordic Journal of Science Fiction and Fantasy Research,* volume 8, no. 1, 2022, pp 7–12.

Chattopadhyay, Bodhisattva. "Speculative Futures of Global South Infrastructures." *Urban Infrastructuring. Sustainable Development Goals Series,* edited by Deljana Iossifova, Alexandros Gasparatos, Stylianos Zavos, Yahya Gamal, Yin Long, Springer, 2022, pp. 297–308.

Colombi, Beatriz. *Diccionario de términos críticos de la literatura y la cultura en América Latina.* CLACSO, 2021.

Csicsery-Ronay, Istvan. *The Seven Beauties of Science Fiction.* Wesleyan UP, 2008.

Davies, William. "Introduction to Economic Science Fictions." *Economic Science Fictions,* edited by William Davies. Goldsmiths, 2018, pp. 1–28.

Duarte, Eduardo de Assis, Melissa E. Schindler, and Adelaine LaGuardia. "Toward a Concept of Afro-Brazilian Literature." *Obsidian,* vol. 13, no. 1, 2012, pp. 97–122.

Eshun, Kodwo. "Further Considerations on Afrofuturism." *CR: New Centennial Review,* vol. 3, no. 2, 2003, pp. 287–302.

Fisher, Mark. "Foreword." *Economic Science Fictions,* edited by William Davies. Goldsmiths, 2018, pp. xii–xiv.

Ghosh, Amitav. *The Great Derangement.* U of Chicago P, 2017.

Hallegatte, Stéphane, and Brian Walsh. "Natural Disasters, Poverty, and Inequality: New Metrics for Fairer Policies." *The Routledge Handbook of the Political Economic of the Environment,* edited by Éloi Laurent and Klara Zwickl. Routledge, 2021, pp. 111–31.

Jasanoff, Sheila. "Future Imperfect: Science, Technology, and the Imaginations of Modernity." *Dreamscapes of Modernity: Sociotechnical Imaginaries and the Fabrication of Power,* edited by Sheila Jasanoff and Sang-Hyun Kim. U of Chicago P, 2015, pp. 1–33.

Jasanoff, Sheila, and Sang-Hyun Kim. "Containing the Atom: Sociotechnical Imaginaries and Nuclear Power in the United States and South Korea." *Minerva* 47, 2009, pp. 119–46.

Jones, Esther L. "African-Brazilian Science Fiction: Aline França's A Mulher De Aleduma." *Obsidian,* vol. 13, no. 1, 2012, pp. 15–36.

Kurlat Ares, Silvia, and Ezequiel de Rosso. *La ciencia ficción en América Latina. Crítica. Teoría. Historia.* Peter Lang, 2021.

Lavender III, Isiah. *Afrofuturism Rising: The Literary Prehistory of a Movement.* Ohio State UP, 2019.

López-Pellisa, Teresa, and Silvia G. Kurlat Ares. *Historia De La Ciencia Ficción Latinoamericana I: Desde Los Orígenes Hasta La Modernidad.* Iberoamericana, 2020.

Medina, Eden. *Cybernetic Revolutionaries: Technology and Politics in Allende's Chile.* MIT Press, 2011.

Melo, Carla. "Urgent (Anti)Spectacles of Critical Hope." *The Utopian Impulse in Latin America,* edited by Kim Beauchesne and Alessandra Santos. Palgrave Macmillan, 2011, pp. 259–74.

Milner, Andrew. *Tenses of Imagination: Raymond Williams on Science Fiction, Utopia, and Dystopia.* Peter Lang, 2010.

Milner, Andrew. "Utopia and Science Fiction in Raymond Williams." *Science Fiction Studies,* vol. 30, no. 2, 2003, pp. 199–216.

Milner, Andrew, and J. R. Burgmann. *Science Fiction and Climate Change: A Sociological Approach.* Liverpool UP, 2020.

Neyrat, Frédéric. "The Black Angel of History." *Angelaki,* vol. 25, no. 4, 2020, pp. 120–34.

Nijhuis, Michelle. "The Tragedy of the Commons is a False and Dangerous Myth: Aeon Essays." *Aeon,* May 4, 2021, https://aeon.co/essays/the-tragedy-of-the-commons-is-a-false-and-dangerous-myth.

Nye, David E. *American Technological Sublime.* MIT Press 1994.

Pratt, Mary Louise. *Planetary Longings.* Duke UP, 2022.

Sanchez-Taylor, Joy. *Diverse Futures: Science Fiction and Authors of Color.* Ohio State UP, 2021.

Schneider-Mayerson, Matthew, Alexa Weik von Mossner, W. P. Malecki, and Frank Hakemulder. "Introduction: Toward an Integrated Approach to Environmental Narratives and Social Change." *Empirical Ecocriticism: Environmental Narratives for Social Change,* ed. Schneider-Mayerson et al., U of Minnesota P, 2023, pp. 1–30.

Seyfert, Peter. "A Glimpse of Hope at the End of the Dystopian Century: The Utopian Dimension of Critical Dystopias." *ILCEA* 30, 2018, pp 1–10.

Shaviro, Steven. *Discognition.* Repeater, 2016.

Shaviro, Steven. "Unpredicting the Future." *Alienocene: Journal of the First Outernational,* April 2018. https://alienocene.com/2018/04/01/futurity-and-science-fiction/.

Srnicek, Nick. *Platform Capitalism.* Wiley, 2016.

Streeby, Shelley. *Imagining the Future of Climate Change: World-making Through Science Fiction and Activism.* U of California P, 2018.

Tironi, Manuel, and Javiera Barandiarán. "Neoliberalism as Political Technology: Expertise, Energy, and Democracy in Chile." *Beyond Imported Magic: Essays on Science, Technology, and Society in Latin America,* edited by Eden Medina, Ivan da Costa Marques and Christina Holmes. MIT Press, 2014, pp.305–29.

Trexler, Adam. *Anthropocene Fictions: The Novel in a Time of Climate Change.* U of Virginia P, 2015.

Vint, Sherryl. "Currencies of Social Organisation: The Future of Money." *Economic Science Fictions,* edited by William Davies. Goldsmiths, 2018, pp. 59–72.

Watkins, Susan. *Contemporary Women's Post-Apocalyptic Fiction.* Palgrave Macmillan, 2020.

Winter, Jerome. "Global Afrofuturist Ecologies." *Literary Afrofuturism in the Twenty-First Century,* edited by Isiah Lavender III and Lisa Yaszek. Ohio State UP, 2020, pp.189–202.

5

Ethnographic and Poetic Interventions

13

Choike Pürrun

The Mapuche People's Sacred Dance

JASMIN BELMAR SHAGULIAN

Ahora es, ahora es, ahora, choyke,
hoy día, hoy día
Porque nos han invitado
Por estas tierras
choyke, habrá baile, habrá baile
por eso ahora aquí estamos,
por estas tierras choyke, choyke
Prepárense, prepáranse choyke
no es la primera vez, será un honor
en esta tierra
hoy día, hoy día choyke, pues choyke
ustedes choyke pues, ustedes choyke
Pasen, pasen choyke, pasen, pasen choyke
Saluden, saluden por estas tierras
Hoy día, hoy día, hoy día pues hoy día
qué lindo, qué lindo
el baile, el baile, el baile
que vienen a mostrar, vienen a mostrar, choyke.
 —Transcription from "Joel Maripil / Choyke Purrun"

The above-cited verses belong to the orality of Mapuche culture, the *ül*. The ül is a poetic verbal expression that could be constructed with a musical melody and then belongs to *ül dungu*. Additionally, those who practice it are called *ülkantufe* (Painequeo Paillán 210). The ül can be performed in different circumstances, from family reunions to community encounters, and serve to entertain, heal, learn, or strengthen relations, among other purposes (210). The ülkantufe develops ideas about ordinary and personal experiences and then composes an ül. This is the case of Joel Maripil, who is a Werken [spokesman] and an ülkantufe from Kechukawin in Southern Chile, near Budi Lake in the IX region of the country. The above text is an ül, composed with a melody to accompany the *choike pürrun*. The ül that Maripil composed belongs to a musical audiovisual performance ("Joel Maripil / CHOYKE PURRUN"), produced by the author, in which the spectator can hear and see how the dance is performed with the help of the traditional musical instruments.

The choike pürrun, the rhea dance, is an important part of Mapuche people's ceremonialism, belonging in the first place to the *nguillatún* ceremony (Silva-Zurita 109), but also to other rituals such as the *kamarikün* ceremony (Cañumil and Ramos 10). The invocation made during the choike pürrun uses the *tayil,* "a powerful ritual song and means of communication generated by the gods. . . . The power of the *tayil* rests in its capacity to attract the auxiliary eagle-spirit [ñamku] to the human domain" (Grebe, "Amerindian" 156).

In the nguillatún, the rogation ceremonial rite, performed by the *machi* (the shaman) and the Mapuche community, invoke nature forces, *ngen*, responsible for the land and the elements whose function is to preserve and give life to the "rivers, mountains, forests, stones, animals, i.e., everything that exists" (Martínez González 43). These natural entities are also beings called upon in the choike pürrun dance. The choike pürrun is a fertility ritualistic zoomorphic dance, "in which the group of people feature as birds . . . that in the ritualistic performance reproduce an activity and a being of a mystical character" (Carrasco 14). In this chapter, I present how the performance is carried out and reflect on its importance in the Mapuche culture. This dance is part of maintaining the equilibrium and the diversity of the cosmos through the sensibilization of the human body. Its purpose is to generate synchronicity with Mother Earth, Ñuke Mapu, and connect with the surroundings by including both humans and other-than-humans, "earth-beings," as de la Cadena denotes them (xviii).

The choike pürrun, as a fertility dance, stands in direct dialogue with the important aspect of the complementary dualism existing in the Mapuche culture—that is to say, the existence of opposing forces of the feminine and masculine in their culture's cosmogony (Grebe, "Presencia" 57) that are endlessly seeking balance. In the Mapuche cosmogony, the *choike* or *ñandú* (rhea bird) has a profound significance for the Mapuche people and their

Figure 13.1. The *choike*'s paws signaling the four cardinal points. Image by Aiyana Shagulian.

ritualistic ceremonies. Montecino indicates that the bird is associated with the ancestors, who, in their plane of existence, go hunting the bird in the Milky Way (487–90). Moreover, the choike is linked to both the lineages in the Mapuche culture and the reproduction, since the dance is a courtship/mating dance performed, for example, during the Nguillatún and We Tripantu (the ceremony of renewal of the cycle) ceremonies (463).

Furthermore, the choike is also a symbolic animal that is connected to the Machi's *kultrún,* the sacred drum of the Mapuche shaman. The instrument has the choike's paws engraved on it (see fig. 13.1), signaling the four cardinal points. In the Mapuche-Tehuelche tale that relates the story of the choike, which has been hunted and trapped, it climbs the rainbow and, as it reaches the sky, tramples the floor hard, giving rise to the *penon choike,* the Southern Cross (Montecino 188–89).

In sum, the choike pürrun, as a ritualistic dance, is part of the Mapuche concept of *Küme Mongen* (good living), defined as a holistic form that encompasses the intimate relation between the well-being of the ecosystems and the humans (Guerrero-Gatica and Achondo 140), which is confirmed by the

Figure 13.2. "Choike Lo Prado 2015" by Ramón Millache. Video still. https://www.youtube.com/watch?v=c6GAqJZmzKQ.

yearly repetition of the enactment of the dance in the most important rituals of the community: the Nguillatún and the We Tripantu. At the same time, the choike pürrun could also be inserted in a wider concept: the cosmological perspectivism or multinaturalism, which Viveiros de Castro explains as the world inhabited by different kinds of subjects, both humans and nonhumans, that perceive reality from a diversity of perspectives from different bodies, not cultures (34). Therefore, the dance, performed by humans, emerges from the knowledge acquired in the existing interrelation between humans and nonhumans (Martínez González 38), and, as Moira Millán points out, the dance is "the Mapuche people's vision on the circular and harmonious relationship with life, which is represented through circular dances" (Poveda Yáñez et al. 83). In the circular movements of this dance, it is perceived as the mystical unity with the forces of nature and mother earth, Ñuke Mapu, that were awakening at the moment of Creation (Robertson 184). Above all, the choike pürrun constitutes a continuity of the traditions of Mapuche people (rural communities) in urban spaces this people inhabits nowadays.

The practical arrangement of the dance incorporates a specific and beforehand-established number of participants and/or ritualistic objects. Grebe explains that "the individuals and objects that fulfill equal or complementary functions are organised in pairs or double couples" ("Presencia" 57). As shown in figure 13.2, the choike pürrun is composed of four choikes or dancers: two women and two men. Nevertheless, the dance was also performed by men originally, since, in accordance with Foerster, the dance denotes the patrilineality of the Mapuche culture, because it is the male choike that takes care of the eggs laid by the female (Foerster 96).

The dance itself around the *rewe*, performed by the dancers, represents the journey (in the nguillatún) through "the cosmos in the space, going through every constellation, always turning around from the Wenu-Mapu, entering Nag-Mapu, and descending as well to the Minche-Mapu" (Huaiquinao 40). The zoomorphic dance relies on four but up to twelve dancers, with a period of rest divided in four. The dance will imitate the mating movements of the rhea bird (50).

The dancers carry a bundle of rattles and plumes (see fig. 13.3) on their heads, or branches of their sacred tree, the cinnamon tree (*Drimys winteri*), and wear full costumes that include "knee-length pants, a belt crossed over the torso with bronze rattles [*kaskawilla*] sewn on it, and a *poncho* over the shoulders" (Poveda Yáñez et al. 85).[1] The dancers move the ponchos up and down, imitating the flapping of the rhea bird's wings, while they move their arms, torso, and feet like the bird. The rattles symbolize the created power, and the suspended movement is associated with the mystic that lies between heaven and earth (Cirlot 124), while the cinnamon branches belong to the symbolic representation of the tree as the life in the cosmos (89), and, more specifically, as a symbol of good, peace, and justice (Montecino 126). The circular counterclockwise movements of the dancers aim to go around the ritual effigy of the *rewe* or the *nillatué* (Grebe, "Presencia" 64; Grebe, "El tayil" 71).

The circular movements are multiple in pair numbers that increase progressively in each of the four sequences of the dance. The dance will start and arrive from the east, which is the auspicious cardinal point into which they direct their pleas (Foerster 59), and the dancers' walk orients them around the effigy. The compasses of the dance are generated through the kultrún and other accompanying instruments that mark the schemes of the successive rhythms (Grebe, "Presencia" 64).

Figure 13.3. "Choike Pürrün familia Porma." Video still. Source: https://www.youtube.com/watch?v=oDwS0RAa7zA

In the context of the nguillatún (act of pleading), the most important religious ceremony, which is celebrated every two to four years and that differs in the diverse communities, the choike pürrun dance conjures fertility in the larger scheme of the nguillatún, where the main purpose is to evoke the divinities, the ancestors, and validate the social order in the community. This is accomplished through animal sacrifice and the performance of other dances besides the choike pürrun, such as the Loncomeo dance (an exclusively male dance), and the Amupurun dance (a female performance). Montecino explains that the modern nguillatún has been adapted nowadays, because it is also conducted in the big cities, especially in the capital, Santiago, and in addition to the traditional meaning it also reproduces the Mapuche identity in big cities, where many Mapuche people live today (464).

The choike pürrun dance represents an interrelation between all living entities in all planes of existence, both material and immaterial, and belongs to the "construction of kinship and their cosmological worldview" (Poveda

Yáñez et al. 84), since most of the time it takes place during the nguillatún, when the lineages are aligned with the supernatural domains, from the Mapu to the Wenumapu (place of the divinities and the ancestors) (Poveda Yáñez et al. 85). In other words, the choike pürrun is a holistic connection that embraces the cosmological consciousness of Mapuche culture, and that represents the social interactions between humans and more-than-humans to strive after a balance in the world.

Note

1 Spanish conquistadors called it *canelo*, cinnamon, because the Mapuche tree had a resemblance to the tree that the Spaniards call "cinnamon" (Montecino 126).

Works Cited

Cañumil, Pablo, and Ana Ramos. "Algunas reflexiones sobre los procesos de formación de lof." *Dinámicas políticas e identitarias de pueblos indígenas: pertenencias, límites y fronteras,* edited by Claudia Briones and Sabine Kradolfer, Mann Verlag, 2016, pp. 1–24.

Carrasco, Iván. "Cruces en literaturas indígenas-mestizas." *Acta literaria,* vol. 54, 2017, pp. 13–27.

Cirlot, Juan. *Diccionario de símbolos.* Ediciones Siruela, 2016.

De la Cadena, Marisol, Helen Risør, and Joseph Feldman. "Aperturas onto-epistémicas: conversaciones con Marisol de la Cadena." *Antípoda. Revista de Antropología y Arqueología,* vol. 1, no. 32, 2018, pp. 159–77.

Foerster, Rolf. *Introducción a la Religiosidad Mapuche.* Editorial Universitaria, 1995.

Grebe, María Ester. "Amerindian music of Chile." *Music in Latin America and the Caribbean: An Encyclopedic History. Volume 1, Performing Beliefs: Indigenous People of South America, Central America, and Mexico,* edited by Malena Kuss, 2004, U of Texas P, pp. 145–53.

Grebe, María Ester. "Presencia del dualismo en la cultura y música mapuche." *Revista musical chilena,* vol. 28, nos. 126–27 (abril–septiembre), 1974, pp. 47 -79.

Grebe, María Ester. "El tayil mapuche, como categoría conceptual y medio de comunicación trascendente." *Inter-American Music Review,* vol. 10, no. 2, 1989, pp. 69–75.

Guerrero-Gatica, M., and P. P. Achondo. "El Bosque y sus Habitantes: Una Discusión Teórico-Metodológica Transdisciplinar del Diálogo Multiespecies." *Revista Etnobiología,* vol. 20, no. 2, 2022, pp. 136–151.

Huaiquinao, Juan. "La cosmovisión y la filosofía Mapuche: Un enfoque del Az-Mapu y del Derecho Consuetudinario en la cultura Mapuche." *Revista de Estudios Criminológicos y Penitenciarios,* no. 6, 2003, pp. 37–58.

"Joel Maripil / Choyke Purrun." YouTube, uploaded by Javiera Gallardo Prosser, 30 May 2012, https://www.youtube.com/watch?v=nTL71pQpmb4.

Martínez González, Omar. "Territorios Danzantes." *Revista Espaço Acadêmico*, vol. 21, no. 231, 2021, pp. 37–48.

Montecino, Sonia. *Mitos de Chile, enciclopedia de series, apariciones y encantos*. Catalonia, 2015.

Painequeo Paillán, Héctor. "Técnicas de composición en el ÜL (canto mapuche)." *Literatura y lingüística*, no. 26, 2012, pp. 205–28.

Poveda Yáñez, Jorge, Beatriz Herrera Corado, and María Mendizábal. "Forced Secularization and Postmodern Discourses within Contemporary Performance: Weaponizing Multicultural Rhetoric to Ratify Asymmetries among Dance Practitioners." *Dance Chronicle*, vol. 45, no. 1, 2022, pp. 79–100.

Robertson, Carol E. 2004. "Fertility ritual." *Music in Latin America and the Caribbean: An Encyclopedic History*, edited by Malena Kuss, 2004, U of Texas P, pp. 182–86.

Silva-Zurita, Javier A. "Music and Cultural Identity: An Ethnically-based Music Program in a Mapuche-Pewenche School." *Eras*, vol. 16, no. 1, 2014, pp. 97–114.

Viveiros de Castro, Eduardo. *The Relative Native. Essays on Indigenous Conceptual Worlds*. Hawoo, 2016.

14

The Paths Not Seen

How I Structured My Representation of Nature

Igor Barreto

English translations by Rowena Hill

I would like to begin these words by remembering a dream: in the mists of the unconscious, I saw myself wandering around the Southern Deserts of Venezuela in the company of a man called Benjamin Cordero; I was helping him to round up a herd of pigs, and, exhausted by the task, we stopped under a lone palm tree to rest, while the pigs rooted in the dark earth. At that moment Benjamin Cordero called my attention to tell me: "When you're going to write a poem do it in an unclean spirit, the dirtiest you can in the world." I have thought often of this episode and the ideas I could associate with the adjective *inmundo,* unclean, which is a term with a deep biblical resonance. The appearances in the New Testament of this unusually forceful adjective are generally linked to the presence of beings seduced by evil, or who represent "evil" as demons, and especially as low-class, marginal creatures. They are representations located on the periphery of a doctrine that, as Christianity did, ended up dominating the West. And, extrapolating this concept to apply it, after the fashion of Baudelaire, to our literary ends, we could say that inmundo is the negation of the focal and also the affirmation of the peripheral.

I would like to mention in my support an essay by the Argentinian critic Josefina Ludmer: "*Literaturas Post-autónomas*" (Post-autonomous Literatures). Also, I will dare in this regard to recall that the inmundo is the contaminated literary expression in various registers from the most diverse expressive forms: what the Argentinian thinker calls "post-autonomous."

Speaking of my poetic writing, it is a concept I have relied on in incorporating narrative elements; stories from life; the highlighting of the properly lexical: especially nouns that are like apparitions of parachronic ghosts, true figures of the past embedded in the present; quotes or paraphrases of texts lost in time. Allow me to add to this expression another that I have always sensed as a kind of impulse favoring the exercise of the cumulative, what I personally call "the force of Implication." For me, poems are constructed by incorporating the greatest quantity of elements that will constitute their horizontality of meaning and atmosphere. The transcendence of any text would thus have to do with its capacity for "implication," with the power of worldly growth the poem can demonstrate. Verticality as a principle of transcendence, of connection with a distant divinity, as happens in the Romantic poems and their successors in the present, has always seemed to me inhuman. Divinity understood in this way lacks what Junichiro Tanizaki calls "the shadow of use," a quality that can only be acquired by the different contacts that "something" has with its surroundings.

In this day and age, contact with the world is summed up more and more often in the construction of an image, an icon, a representation of decadent and neutral manufacture, that refuses to deal with the wealth of the circumstantial profile of the world, hiding behind an often boastful lyricism. In an interview, I read that Cioran asserted that if poetry continued like that, it could expect an operatic—redundant, high-sounding—destiny. I would like to be able to speak to you of a greater complexity, an aspiration involving the richness of each and every one of the linguistic elements, especially those that confer a greater verbal impact on the poem, a "force of gravity" on the words, which I would like to see weigh so heavily that their weight would make them fall, leaving a hollow in the ground.

The words above, together with the final fragment in this chapter, make up the outflow from the passages I am going to reproduce, and that will have as labels or titles the different problems that have attracted my curiosity in the unfolding of my creative process. These quotes are taken from a book I published in 2006 with the title *El Llano Ciego* (The Blind Plain), that I now recall, with memory and soul, as the Spanish poet of the Generation of 1927, Don José Bergamín, would have said. I present these fragments from *The Blind Plain*, under headings that refer to the problems they put forward, and accompanied by poems related to the texts, which give an account of my process and my lament when I think of the contemporary representation of nature.

Exile

Exile as a category of existence, the abyss and
pallor of thought as now internalized conditions
of our daily life. The first form of exile we
suffer is one whose sphere is time. Our present
is only the time of deep disillusionments. That's
the cause of the unreality that haunts us. Very
few of us have a homeland in a country of the past.
Derek Walcott once dared to assert that "amnesia"
was at the origin of the New World. I speak of
these things as a reader of poems, those vessels where
we keep the current of sensibility.
In film there is a compositional device that might be
revealing. I'm referring to the application of the "imaginary
geography" by which a location is constructed adding to
the montage parts or images from different locales. Although
the parts are real the whole is always imaginary. This
happens in the mind of an exile; his memories correspond
to concrete places, but "the whole" of his life is imaginary.

Nature of Exile

(Apartment nocturne, 1998)
Some cattle arrived from a yearning for woods,
from some sorely missed hills.
What was the meaning of those animals
with their human faces?
The kitchen was a bonfire
at midnight.
The vegetable
hush of the balcony
where some ferns
flutter like sphinx moths.
What happened to the quiet of those places
I knew so well?
I didn't find patterned mire
nor the blue shirt.

It was the nature of exile,
a river of nothing.
Something that cuts an onion in small pieces,
white, like a streetlamp under a withered tree.

Paradise Lost

Milton said
in *Paradise Lost*:
*The earth so small
compared to the sky
and without the light.*
So, an earth
in essence dark.
Poor deluded tropics
that believe the light belongs to them.
The palm with shining fronds
has lost its pride
and is sick:
it's only
a relic
of the shade.

Nature

What can the image be that I'm seeking for nature? It's
certainly not one that's deified and spiritualized ad nauseam.
But neither is it that other more modern image that
Gottfried Benn speaks of in his letter to Oelze: *Nature
is empty, deserted; only the deer see something in it,
poor devils condemned to constant suffering. Flee
from nature! It spoils thoughts and noticeably impairs
style!* Benn, whose voice comes from the urban netherworld,
expresses a vision where pity is absent. The German poet
doesn't want to be the voice of a collective, nor of the bourgeois
imposture of its values: solidarity, authenticity, identity,
transcendence. Although I tend to admire his pitiless, shorn,
free poetry, I can't stand his detachment and vacancy of soul.
If nature has been perverted and invaded and survives if at all

in the imaginary of exile, we are obliged to take on this
awareness of degradation (and I'm remembering James Hillman),
and prepare to restore soul to the world.
Nature? The nonhuman, simply that
portion of the cosmos I haven't seen, the most remote, the place
where some birds eat seeds of trees for which
I lack names. The mere presence of a person frightens away
nature, unsettles it with its pride and humanity. Saying
"man" and fleeing, everything that happens without any law,
is the same thing.

The Present

I REMEMBER a walk with the chronicler of the town of San
Fernando, Don Julio César Sánchez Olivo. We stopped at
each corner and he would tell me: "Here was the 'Liberty
Pharmacy'" (now there was a building); "Here was the old
ice factory" (now there was only a vacant lot). After those
walks, I thought that every object deserved to endure and be
a reminder of some time, since only what's old has heart.
St. Augustine believed that the present should be conjugated as
past-present or future-present. But, lamentably among us,
lamentably for things, for streets and cities, here the
present follows the present in a world of pure, dense
daily triviality.
The city that the Conquistadors built was a walled
city (a fort city), so different and similar to
the contemporary city, also walled, but by the
present, the wall of the present. From there its terrible
insularity derives. Paraphrasing the Cuban poet
Virgilio Piñera, we could say: the accursed circumstance
of the present everywhere.

Ungaretti

I HEARD Ungaretti speak
about his Alexandria,
shut his blue eyes and say
other places in the Orient

may have their thousand and one nights,
but Alexandria has a desert.
We also have one:
the amnesia and the desert of the present.

Landscape

The landscape has died. The landscape of Romantic tradition has
died, although we still discover traces of the glorifying lyric
in our poems. In that convention nature was identified with an idyllic
state before the "fall." Nature was its "paradise lost," something
that deserted us when we passed through the gates of childhood,
as we read in Wordsworth's "Ode on the Intimations of Immortality
from Recollections of Early Childhood." We were happy like
Rousseau within the bounds of that landscape until the arrival
of modernity, which hastened our abandonment of our ancestral and
collective memory. But modernity also brought ideological and
linguistic awareness, pointing to the great load of mere props
accumulated in our vision of nature. Though it has to be said that
Romanticism was driven by a national spirit of recognition of
geography, where the representation of the landscape became a way
to embody aesthetically what in other spheres was a political destiny.
The rule of the landscape is monotheism. This erroneous
perspective arises at the apex, the seat of the eye (one and
deified) that segments nature. But we need only observe the
cosmos with its names: Ceres, Venus, Neptune, to realize
how many are watching us from the tops of trees and
the swell of steppes.
The landscape has got trapped in a compulsive desire to
idealize. It survives in crystallized images. At its limits
nature is God, outer or inner according to whether we think
of Plato or St. Augustine. The territorial exile out of which I speak
still retains an atom of reality, an objective correlative,
as Eliot would say, the memory of an experience which
claims to be historical.
Where are the revered ruins of nature if today what
we find is the rubble of a river of shit? How can we
go on believing in the landscape as a beautiful and pleasant

representation? The contemporary landscape (if we insist on that term) would be a corrupted, invaded, impure representation: a cordillera of garbage. How can we jump by means of a lyrical strategy over this present and go back to writing about trees that nod their heads and murmur sweet nothings among themselves?

Place

If someone said they wanted to represent the landscape it would reclaim for poetic writing the notion of "place." "Place" is dynamic as opposed to the static character of a pictorial image of nature: particular, historical as against universal, nominal as against adjectival. Landscaping? According to Baudelaire it consists only of glorifying vegetables.
Léon Bloy said: "The lowest degree of wretchedness is, surely, not having what can be called a home."
To confront evil, I have no deities; I have the memory of places. That's what I can call on to help me.
The soul exists only as a relation between the individual and "place." You have a soul when you are (harmoniously) in any locality however remote. So that you encounter again the possible unity of place where the soul becomes palpable to the senses. I think of these ideas while I look at Giotto's *Joachim Among the Shepherds,* one of his frescoes in the Arena Chapel in Padua. Giotto discovered the notion of place when perspective was merely a hint, and human figures and nature were a modest presence, their frontality undisguised. Between Joachim returning repentant and the shepherds receiving him together with sheep, rocks and trees, a spiritual community is created, a sympatheia that values space, place, beyond any naturalistic perception.
What a paradox:
a seal oil lamp lights up the face of Robert Flaherty, the clever founder of documentary film. It was the second decade of the twentieth century and he had decided to abandon his profession as a mineralogist. On his last journey to Baffin Land (north of Canada), he discovered the dominance of

the human where the cycles of the inhospitable subarctic
steppes rotate in the boreal character of the Eskimos.
After two months in canoe and sled, Flaherty reached
them. He had with him a considerable load: two Akeley
movie cameras, the best in glacial temperatures, since they
use a minimum of oil and grease; also, a tripod for the
gyroscopic cameras; dozens of tins of virgin film
from Eastman Kodak; and a complete laboratory that
allowed him to develop the film he was using . . . And all
that (what a paradox!) to find himself in an objective
and convincing way in contact with nature.

The Past and Memory

Go beyond reality, go beyond the landscape. Go deep into
Memory, which is pure verbal creation.
Memory is a spontaneous text that comes toward the present
and our consciousness intercepts its passage. Consciousness
calls to memory from the shore of verbal nakedness.
Remembering means invoking the words of a prayer, commending
oneself to the spirit of a place and to an occurrence.
Speaking of memory means entering a binding space.
I mean that the people and objects called up relate to
form a plural network, a plural identity, where the "I"
must inevitably draw back.
In the sphere of memory, the present accommodates itself to the past,
bowing to it.
There are fatal correspondences between Akira Kurosawa's film
Dersu Uzala (1975) and the tragedy of the head buttons in
Friedrich Murnau's classic expressionist film *The Last Man*
(1924). Murnau's character (played by Emil Jannings) saw
his identity, his name, the doorway to his selfhood tragically
supplanted by a supposed social identity given him by the
flamboyant uniform of the Hotel Atlantic. "Pangermanic"
Germany, where Fascism and Communism germinated,
introduced in our century these mutilations of the person. In the
spirit of the theater of the "I," which the Expressionists called

ich-drama, the character played by Jennings is led to a burlesque
and dishonorable end. As for Dersu, old and trapped by a fierce
deity that lies in wait for him in the shape of a Bengal tiger,
he decided to give up everything that made him admirable:
his nature as a man immersed in the jungle between Russia and
China, and his shrewd hunter's culture. His panic led him to
take refuge in the small city house of the Arseniev family,
far from the taiga. Dersu defied his destiny and when he tried
to go back, almost blind, an ignominious death surprised him at
the hands of some thieves. When someone deviates from tradition,
when they lose the name, the title, of the story we uniquely are,
a banal end (almost always) follows, and body and spirit meet
a faded death.

Some Proposals

The life of a man goes by building up, refining, one or
more histories. Stories where the narrator sums up the keys
to his existence, his relation to nature, people, and
things. I heard a fisherman tell how his brother died
drowning in a river, and connect that fatal hour with the
cry of the curassow hidden in the gallery forest. For him
it was the voice of loneliness and silence. These stories
powerfully develop deep realities. They refer obliquely to
the intimate world of the teller, his interests and concerns:
these are the *essential histories.* I look for them, I discover
them, and I work them up as poems.
From the beginning I felt a desire to give my verses more
substantiveness. The first device I recurred to was the image.
Organizing the poem by means of a *constructive montage*
in the manner of Pudovkin, where the ordering of a series of
takes composes the strophes, and so on, sequence after
sequence, to the end of the text. It was only visual engineering.
That mode favored the sense of sight and contained in its figurative
design the germ of its own destruction: the poem and the
word lost resonance and gained an excess of rationality.
It was then that the image came to my mind of a fisherman

on the bank, hidden in a bend of the river in the gallery forest,
watching without missing a detail the reflecting surface of
the water. Watching and, while watching, applying all the intensity
of a person listening with extreme attention. There you have
the answer (I said to myself): *look like a person listening.*
Relating sight with that sense, with hearing, that for
St. John of the Cross was the most spiritual of all. In that
way, the world represented in the poem acquired greater
depth and its image resonated with human emotion.

The Centaur

Tied to a rope
I led the centaur
to the shed
at the back of the house.
You were the wise
master of Achilles
and of Aesculapius,
and with one slash
I cut open his head
giving him
a muzzle
with his thick lips.
I whispered in his ear:
The savannah is the void
where the horse
is all that exists.
Vulgar horsemen
will come
to steal
your transcendence.
At the end
expect sadness,
evil
and defeat.

Pastoral

PASTOR Caeiro
do not kill
lyrical poetry.
You think
of the poem
and you leave us
in the wilderness
with the complaint
of some amnesiac
sheep
who only bleat
symbolically.
If you could see
what became
of Titirus and Salicius:
from the Latin garden
to the shed
made of three planks
at the thickening
edge
of the city.
I've spoken to them:
Where there was
green
they can't paint
with green!
But they
pay no attention:
if the wind blows
it's because the air
is lamenting,
and turning off the tap
they hear the sound
of a voice . . .
But it's not their fault.
Our problem is in our souls,
Horace said.

Last Fragment

I have always thought that the only possibility the earth had of saving itself from our predatory spirit would be our being faithful to an ethic that considered *Nature* as an other. This attitude would imply that there must exist a *distance* characterized by the awakening of compassion, and then, later, would come the recognition or learning of some forgotten norms. This *distance* would have to be religious in the sense that we must follow the movement of nature, recognizing it as a superior being, as happens in the Prayer of the Heliotrope, quoted by the late Neoplatonic philosopher Proclus, in his treaty on Greek hieratic art. It has seemed to me that starting from this prayer, emulating it, we could look at nature with the same reverent *distance*. This idea is largely inspired by the writings of Henry Corbin, and more especially his book on creative imagination in Ibn Arabi. I believe this principle of a necessary *distance* could be the basis of a possible ethics of earthly and worldly coexistence for people of today, dominated by a desire for destructive possession which has put an end to the necessary sympathy with our world, the correspondence between the stages of our lives and those of natural processes, or the interlocution that can only be born of this understanding.

CONTRIBUTORS

Emily Baker is Senior Lecturer of Comparative Literature and Latin American Studies at University College London (UCL). She is the author of *Nazism, the Second World War and the Holocaust in Contemporary Latin American Fiction* (Cambridge UP, 2022).

Igor Barreto is a Venezuelan poet and essayist. Recent publications include the poetry collections *The Blind Plain* (Tavern, 2018) and *La sombra del apostador* (Visor, 2021).

Ken Benson is Emeritus Professor of Spanish at Stockholm University. He is the author of *La subversión silente. Carmen Laforet: poética y hermenéutica* (Albatros, 2024).

Jasmin Belmar Shagulian is Lecturer of Spanish at Luleå Tekniska Universitet.

Patrick Brock is a writer and researcher based in Brazil who studies the intersection of speculative fiction, futures, and activist practices in Latin America. He obtained his PhD from the University of Oslo (2024).

Azucena Castro is Assistant Professor of Latin American Cultural Studies at Rice University. She is the author of *Posnaturalezas poéticas: Pensamiento ecológico y políticas de la extrañeza en la poesía latinoamericana contemporánea* (De Gruyter, 2025) and editor of *Futuros multiespecie: Prácticas vinculantes para un planeta en emergencia* (Bartlebooth, 2023).

José Carlos Díaz Zanelli is Visiting Assistant Professor of Hispanic Studies at Hamilton College. He is the author of *Insurgent Veins: Indigenismo, Indigenous Literatures, and Decolonial Cracks* (U of Pittsburgh P, 2025) and coeditor of *Worlding Latin America: Corpus, Praxis, and Global Networks* (De Gruyter, 2024).

Allison Mackey is Professor of English Literature at Universidad de la República, Uruguay, and Research Associate at the University of the Free State, South Africa.

Montserrat Madariaga-Caro is Assistant Professor of Hispanic Studies at Vassar College.

Paul R. Merchant is Associate Professor of Latin American Film and Visual Culture at the University of Bristol. He is the author of *Remaking Home: Domestic Spaces in Argentine and Chilean Film, 2005–2015* (U of Pittsburgh P, 2022) and coeditor of *Latin American Culture and the Limits of the Human* (UP of Florida, 2020).

Andrés Ernesto Obando is Lecturer of Hispanic Literature at Uniminuto Institution, Colombia, and research affiliate at Grupo de Investigación en Humanidades Ecológicas (GHECO), Universidad Autónoma de Madrid.

Roberto Robalinho is Postdoctoral Researcher at Universidade Federal Fluminense and the University of Tübingen.

Victoria Saramago is Associate Professor of Hispanic and Luso-Brazilian Studies at the University of Chicago. She is the author of *Fictional Environments: Mimesis, Deforestation, and Development in Latin America* (Northwestern UP, 2020) and coeditor of *Handbook of Latin American Environmental Aesthetics* (De Gruyter, 2023).

Gianfranco Selgas is British Academy Postdoctoral Fellow at University College London (UCL). He is the author of *Regionalismo ensamblado: Cultura, ecología política y extractivismos en Latinoamérica (1930–1940)* (Iberoamericana Vervuert, 2025) and coeditor of *Energy Matters: Latin America and the Cultural Critique of Energy* (Environmental Humanities, 2025).

Sebastian Wiedemann is Assistant Professor of Film Studies at Universidad Nacional de Colombia. He is the author of *Deep Blue: Future Memories of a Living's Cinematic In-Between* (Evidence, 2019) and coeditor of *Migrant Thoughts: Cinematographic Intersections* (Hambre, 2020).

INDEX

Page numbers in *italics* refer to illustrations.

Aboaf, Claudia, 218
Abyss (film), 202–9
Acilde (protagonist). See *La Mucama de Omicunlé* (Indiana)
actants, 78–79, 82, 89n5, 140
Acuña, Máxima, 15, 76–79, 83–85, 87
aesthetics, role of, 5
AF. *See* architectural fiction
Afro-Brazilian literature, traits of, 223–24
Afrofuturism. *See* Brazilian Afrofuturism, climate fiction and
Afro hair, empowerment of, 230n6
After the End: Representations of Post-Apocalypse (Berger), 70
Agosín, Esteban, 52–53
Ain-Zaila, Lu, 216
Alaimo, Stacy, 47
Alampi, Antonia, 163
Alas de mar (film), 42, 51
Albert, Bruce, 147, 209
Alenso, Ana, 94–95, 106–8
Alice Springs, ghost gum trees in, 150–53
Alimonda, Héctor, 85
Allende, Salvador, 230n7
alternative futures, projecting, 45–49
aluminum, 95, 103, 195
Amazon Environment Research Institute (IPAM), 144
Amazonian Indigenous cosmovision, 106
Amazon (rainforest), 4, 7, 76, 83, 86, 116, 144, 156n3, 201
Amupurun dance (female performance). *See choike pürrun* (Mapuche dance)
Anane, Mike, 161
Anarquistas del Norte, 173

Andes, 4
Anthropocene, 5, 11, 16, 79, 184, 198; acknowledging Indigenous scholars, 184; advocating for term use, 198; "Anthropocene fictions," 141n2; climate novels and, 216–17; colliding two temporalities of, 150; embodying awareness of, 127–43; ghosts of haunted landscapes of, 163, 172, 178n9; and Ozymandias effect, 70; and post-catastrophic landscapes, 57–61; relations between image and, 144–60; visualizing, 146–48. *See also* awareness (of Anthropocene), embodying
Anthropocene Islands, 50
anthropology, "speculative turn" in, 47
Antillanca, Cristian, 27
Antofagasta, Chile, deterritorialization of, 96–101
Antofagasta, mining areas of, 95
Antonioni, Michelangelo, 149
Aparicio, Juan Ricardo, 80
Arboleda, Martín, 93
archipelagic thought, 49–53
Archipiélago (multimedia project), 42, 52–53
architectural fiction (AF), 226
Ares, Kurlat, 218, 224
Argenis (character). See *La Mucama de Omicunlé* (Indiana)
Argentina, 2, 25, 127, 141n3
Arica, Chile, portraying. See *Arica* (film)
"Arica: A Toxic Waste Scandal with Our Soil Panel" (seminar), 178n7
Arica (film), 162–63, 177n5; aerial view repetition in, *166,* 166–67, *167;* contaminating modern narrative of progress, 168; context of, 165–66; experimental footage in, 167–68; juxtaposition of archival images with

present-day footage, 168; opening scene of, 166; overview, 176–77; perceiving North-South relations, 168; placing toxicity as external and internal phenomenon, 168–69; spotlighting national governments, 170; and subclinical toxicity, 169; unraveling ghostly presences, 167

Article 19 (Section 8) (Chilean constitution), 44–45

artistic forms, political ecology of, 73; archives of the planetary mine, 92–110; ecological conflicts in Peru, 75–91; fiction writing and environmental conservation, 111–24

artistic interventions (in Chile): archipelagic thought, 49–53; constitutional derangement, 42–45; overview, 41–42; projecting alternative futures, 45–49

"artistic operations," 27

artivism, 9, 13–14, 45

Arts of Living in a Damaged Planet, The (Tsing), 60

Asháninka (people), 80–83

Atahualpa (Inca king), 84

audiovisual composites, 210

Australia: ghost gum trees in, 150–53; wildfires of 2019/20, 153–55

awareness (of Anthropocene), embodying: caring for "monstrous" children, 129–30; ecogothic lens, 129; flexibility of gothic, 128; homaging Lovecraft, 130–34; nature/landscape as source of fear, 128–29; overview, 127–28, 140; questioning limits of monstrosity, 134–40; resource gothic, 129

AzMapu, 13–14, 26–30, 32–34, 36n9

Azócar, Cristóbal, 50

Baixada Fluminense (Rio de Janeiro), 219

Bajo el agua negra (Enríquez), 127; betraying cautious and ambivalent sense of possibility, 139; ecocentric awareness in, 133; as "mixed bag" tribute to Lovecraft, 131–32; nature/landscape as source of fear in, 128–29; "pacts of indifference" in, 130; paying attention to register of super/natural horror, 132–33; protagonist of, 130; reemergence of nonhuman life in, 133–34; social criticism in, 131

Baker, Emily, 183
Barbas-Rhoden, Laura, 128
Barbosa, Elson, 122n5
Barreto, Igor, 2–3, 245
Barros, María José, 46
Battarbee, Rex, 151
Bazterrica, Agustina, 133, 135
Beasley-Murray, John, 59, 63, 64–65
Beautiful Soul Syndrome, 189
Benjamin, Walter, 155
Bennett, Jane, 140
Benson, Ken, 1
Bergamín, Don José, 246
Berger, James, 70
Berrospi, Francisco, 81–82
Bingham, Hiram, 59–63, 67, 68
Binns, Niall, 88
biocoloniality, 85
Black middle class, rise of, 217
Black Summer, coverage of, 153–55
Blaser, Mario, 77, 79, 80
Blind Plain, The (Barreto), 246; "Exile," 247; "Landscape," 250–51; "Last Fragment," 256; "Nature," 248–49; "Nature of Exile," 247–48; "Paradise Lost," 248; "Pastoral," 255; "Place," 251–52; "Some Proposals," 253–54; "The Centaur," 254; "The Past and Memory," 252–53; "The Present," 249; "Ungaretti," 249–50
Blybarnen (film), 165–66
Boal, Augusto, 43
Bobbette, Adam, 92–93
Boliden-Arica, invisible toxic link of. See *Arica* (film)
Boliden (company), 165
Bolívar, mining areas of, 95
Bolivian Highlands, 17, 162–63, 170
Bolivian Railroad Company, 174
Bolle, Willi, 122n3
Bollington, Lucy, 8
Book Biennial, 219
Borum du Watu (people), 149–50
Bould, Mark, 5
Bozcaada International Festival of Ecological Documentary, 177n5
Braidotti, Rosi, 10, 127
Brasil 2048 (Ain-Zaila), 216, 220, 227, 230n4

Brazil, 16, 18, 155–56; fixing monopoly on possible futures, 219–20; *Grande sertão: Veredas* as environmental agent of, 113–15; heuristic imagination in, 225–27; Mariana dam disaster in, 145, 149–50; postapocalyptic nation-building in, 221–23; refoundation of, 216–17; trauma and collective action, 223–25

Brazilian Afrofuturism, climate fiction and: fixing monopoly on possible futures, 217–20; heuristic imagination, 225–27; overview, 216–17; postapocalyptic nation-building, 220–23; shared intelligence, 227–30; trauma and collective action, 223–25

Brief History of Neoliberalism, A (Harvey), 218

Brock, Patrick, 216
Brown, Kendall, 93, 100
Buell, Lawrence, 169, 177n1
Bullard, Robert, 162, 177n1
Burges, Joel, 217
Butler, Gavin, 154
Butler, Octavia, 239
Byron, Glennis, 127

Cadaver exquisito (Bazterrica), 135
Cadena, Marisol de la, 47
Cádiz, Ilda, 218
Cajigas-Rotundo, Camilo, 85
Caldera, Rafael, 102
Calderón, Martín González, 50
Canaima National Park, 115–18
Canepa, Andrea, 94
capitalism, role of geology in, 92
Capitalocene, 4, 16, 92, 109n2, 185, 198
care, denouncing fundamental failure of, 136
Carga sellada. See *Sealed Cargo* (film)
Carpentier, Alejo, 111, 115–18, 184
Carrigan, Anthony, 7
Carson, Rachel, 177
Cartier-Bresson, Henri, 150, 157n10
Castellano, Carlos Garrido, 183
Castro, Azucena, 1, 161
Castro, Eduardo Viveiros de, 47, 81
Castro, Viveiros de, 148
Castro-Klarén, Sandra, 59
Catrillanca, Camilo, 33
Caycedo, Carolina, 94

Chandler, David, 47, 50
Chattopadhyay, Bodhisattva, 226–27
Chávez, Hugo, 102
CHECO. See Chilean Economic Simulator
Chile, 2, 9, 14–17, 93–94, 107; archaeological heritage, 34; archipelagic thought in, 49–53; artistic interventions in, 41–56; collective action in, 225; constitutional derangement in, 42–45; copper extraction in, 96–101; death by *terricidio* in, 32–34; "ecological avant-garde," 45–49; LasTesis performance in, 41–42; micropolitics of life in, 25–40; poetic portrayal of death in, 30–32; projecting alternative futures, 45–49; "sacrifice zones" in, 45; as settler state, 28–30; unequal toxic waste sender-receiver relations in, 165–70; "usurpation law" in, 29. See also *Arica* (film); Mapuche poetics

Chilean Economic Simulator (CHECO), 230n7

China, 162

choike pürrun (Mapuche dance): circular movements of, 241; as fertility ritualistic zoomorphic dance, 238–39; importance of, 238; and *Küme Mongen* concept, 239–41; nguillatún context, 242; performing, 241; practical arrangement of, 241; representing interrelation of all living entities, 242–43; *ül* accompanying, 237–38

choike (rhea bird), 238–41

Chota, Edwin (character), 15, 76–80, 82–83, 85–87. See also *Guerras del Interior* (Zárate)

CineEco Seia, 177n5

cinematic practices, notes for ecology of, 201; acting in spirit, 209; audiovisual composites, 210; bifurcation of nature, 202–3; cinema of process, 203–4; considering atmosphere, 211; creation of partial connections, 211; decolonizing cinema, 204–6; expressing otherness, 207–8; formula proposal, 202; genetic potency of a world without an image, 202; gestures of care and healing, 211–12; intrusions of Gaia, 211; openness to constant destabilizing and hesitating action, 210–11; preoccupation/obligation of cosmopolitics, 213; search for minimal events, 206–7; utupë, 209, 212–13

262 · Index

Cioran, Emil, 246
climate crisis, 1, 2, 6, 14, 25, 35, 41, 44, 87
climate fiction: fixing monopoly on possible futures, 217–20; heuristic imagination, 225–27; overview, 216–17; postapocalyptic nation-building, 220–23; shared intelligence, 227–30; trauma and collective action, 223–25
climate unthinkable, contesting: archives of the planetary mine, 92–110; artistic interventions in Chile, 41–56; Brazilian Afrofuturism, 216–34; cinematic portrayal of the dumping of toxic waste, 161–79; cosmopolitics of the image, 201–15; depictions of postcatastrophic landscapes, 57–72; ecological conflicts in Peru, 75–91; ethics in *La Mucama de Omicunlé*, 183–200; ethnographic intervention, 237–44; fiction writing and environmental conservation, 111–24; Mapuche poetics, 25–40; overview, 1–4; photographing aftermath of natural disasters, 144–60; poetic intervention, 245–56; resisting and imagining from the South, 4–9; Río de la Plata gothic, 127–43; understanding, 9–13
coevalness, denial of, 119
Colchester, Marcus, 112
collective action, climate fiction and, 223–25
collective environmental memory, preserving, 115–18
Colombia, 2, 33
coloniality, reproduction of, 84–85
coltan, 94, 101–2, 105
constitución ecológica (ecological constitution), 48
Constitutional Convention (in Chile), road to: archipelagic thought, 49–53; constitutional derangement, 42–45; overview, 41–42; projecting alternative futures, 45–49
constitutional process, need for, 43–44
copper, 95–96
Cordero, Benjamin, 3, 245
Coronil, Fernando, 105, 107
Corporación Traitraico, 46
corporeality, mutability of, 47
corporeality (of Mapu), nurturing, 30–32
cosmic events, potencies of, 208

cosmopolitics of the image. *See* cinematic practices, notes for ecology of
Cossio, José Gabriel, 67
Cousteau, Jacques, 186–87
Cronon, William, 121n2
Crutzen, Paul J., 144
Cthulhu, evoking. *See Bajo el agua negra* (Enríquez)
Cuban Revolution, 192–93
cultural appropriation, term, 151–52
cultural crisis. *See* gothic, engagement with
cultural expressions: archives of the planetary mine, 92–110; artistic interventions in Chile, 41–56; Brazilian Afrofuturism, 216–34; cinematic portrayal of the dumping of toxic waste, 161–79; cosmopolitics of the image, 201–15; depictions of postcatastrophic landscapes, 57–72; ethics in *La Mucama de Omicunlé*, 183–200; ethnographic intervention, 237–44; fiction writing and environmental conservation, 111–24; limits of culture, 75–91; Mapuche poetics, 25–40; overview, 1–4; photographing aftermath of natural disasters, 144–60; poetic intervention, 245–56; resisting and imagining from the South, 4–9; Río de la Plata gothic, 127–43; understanding climate unthinkable, 9–13
culturalist explanations, limits of, 77–79
culture, limits of: anthropology of ontologies, 80–83; decolonial scopes of Indigenous activisms, 83–87; limits of culturalist explanations in, 77–79; overview, 75–77, 88
Cuñachí, Osmán, 76–77, 85, 87
Cusicanqui, Silvia Rivera, 36n8
Cuyanao, Jaime, 26
Cyberpunk 2077 (video game), 229–30

Davies, William, 218
Davis, Heather, 184
dead, reclaiming, 32–34
deadly affairs, 162–65
death, poetic portrayal of, 30–32
de Castro, Viveiros, 157n5–6, 240
Deckard, Sharae, 184
Decolonial Ecology: Thinking from the Caribbean World (Ferdinand), 184

decolonization (of cinema), 204–6
deep temporality, 101–7
de la Cadena, Marisol, 151
Deleuze, Gilles, 202
Delight Lab, 45–49
DeLoughrey, Elizabeth, 7
Del Príncipe, David, 129
Demos, T. J., 94
derangement. *See* artistic interventions (in Chile)
Derrida, Jacques, 60, 178n9
Desviar la inercia (film), 94; highlighting "deterritorialization" of Antofagasta, 99–100; historical trend in Chilean mining zones, 100–101; industrial processing sounds in, 97–98; mechanical mentioning of names in, 99; overview, 107–8; plot of, 96; relationship between invisible and auditory, 97–99; subsoil visibility in, 96–97
deterritorialization, 96–101
Didur, Jill, 7
"disaster frame," employing, 57–58
Diverting inertia. See *Desviar la inercia* (film)
Donovan, Amy, 92–93
Duarte, Eduardo de Assis, 223–24
Dumping in Dixie (film), 177n1
dumping of toxic waste, cinematic portrayal of: *Arica,* 165–70; deadly affairs, 163; environmental racism, 162–65; ghost acres, 163; national relations, 170–76; North-South relations, 165–70; overview, 161–62, 176–77; unequal sender-receiver relations, 165–70
Dwyer, John, 61

earth, archives of. *See* planetary mine, archives of
Earth, temporality of, 104–5
"earth-beings," 151. *See also* ghost gum trees, destruction of
Echevarría, Esteban, 133
Echevarría, González, 117
ecocriticism, 77, 88, 129, 229
"ecological avant-garde," 45–49
ecological conflict. *See* Peru, ecological conflicts in
ecological constitution. *See* artistic interventions (in Chile)

ecological zones, vertical control of, 68
Ecology Without Nature (Morton), 187
economy: of global waste, 162, 176; political economy, 104, 218, 230n4
economy of the image, term. *See* environmental disasters, photographing aftermath of
Edman, Lars, 162, 164
El botón de nácar (film), 50
El fin de la raza (Herrera), 218
Elias, Amy J., 217
Ellena, Nicole, 48
El matadero (Echevarría), 133
El Ojo y La Flor (Aboaf), 218
El Rey del Agua (Aboaf), 218
Embry, K., 131
Ena Dias da Silva (protagonist). *See (In) Verdades: Uma heroína negra mudará tudo* (Ain-Zaila)
Enqvist, Gunnar, 177n6
Enríquez, Mariana, 17, 127–33, 141n4. See also *Bajo el agua negra* (Enríquez)
environmental catastrophism, contesting, 181; Brazilian Afrofuturism, 216–34; cosmopolitics of the image, 201–15; ethics in *La Mucama de Omicunlé,* 183–200
environmental conservation, fiction writing and: fictional universe shaping primary universe, 119–20; overview, 111–12, 120–21; preserving collective environmental memory, 115–18; referencing national parks in fiction, 112–15
environmental destruction, visibility of, 125; cinematic portrayal of the dumping of toxic waste, 161–79; photographing aftermath of natural disasters, 144–60; Río de la Plata gothic, 127–43
environmental disasters, photographing aftermath of: Anthropocene visualization, 146–48; Australian wildfires of 2019/20, 153–55; ghost gum tree destruction, 150–53; Mariana dam disaster, 149–50; overview, 144–46, 155–56
environmental politics and culture, critiquing, 196
environmental racism, 17, 162–65, 171, 172
Epeli Hauʻofa, 49
escapism, critiquing, 185–87

Espíritu del agua (animations), 46–49
estallido social (social uprising). *See* artistic interventions (in Chile)
Europe, ruins in, 61
evil, imaginary of, 80. *See also* Asháninka (people)
exploitation, 8, 11–13, 175, 217, 227; of cheap natures, 185, 194; Indigenous bodies and, 34, 172; planetary mine and, 92–95, 99–102, 107, 109; plantation system and, 146; struggle within legacy of, 83–87
externalization (of nature), 64–67
extraction, materiality of, 101–7
extractive zone, 94

Fabian, Johannes, 119
Falling Sky, The (Kopenawa), 147–48, 209
"Faumelisa Manquepillán—Poema La Materia—(Wetruwe Mapuche)" (video), 26
"felt theory," 36n8
female protagonists, witnessing contemporary moment through: caring for "monstrous" children, 129–30: ecogothic lens, 129; flexibility of gothic, 128; homaging Lovecraft, 130–34; nature/landscape as source of fear and terror, 128–29; overview, 127–28, 140; questioning limits of monstrosity, 134–40; resource gothic, 129
Ferdinand, Malcom, 184
Ferrante, Lucio, 131
FF. *See* future fiction
fiction writing, environmental conservation and: fictional universe shaping primary universe, 119–20; overview, 111–12, 120–21; preserving collective environmental memory, 115–18; referencing national parks in fiction, 112–15
Figueroa, Sebastián, 184
Fillol, Alberto Serrano, 50
Final Plan, 119. *See* Canaima National Park
Fisher, Mark, 60, 219
fog, Río de la Plata region and, 137–38
Fornoff, Carolyn, 10
fossil fuels, 1, 4, 92, 107, 144
French, Jennifer, 88
Fritsch, Kelly, 169
future fiction (FF), 226

Gaia, intrusions of, 211
Galindo, Flores, 59
Gallegos, Rómulo, 123n11
Gan, Elaine, 163, 167
garbage imperialism, term, 163, 177n4
Gell, Alfred, 145–46
geology. *See* planetary mine, archives of
Ghana, 161
Ghosh, Amitav, 4–6, 7, 44, 133, 219
ghost acres, 163, 167, 177n3
Ghost Gum (paintings), 152–53
ghost gum trees, destruction of, 150–53, 158n11
ghostly double gazes, configuring, 17, 164, 168–69, 170, 176
Giffard-Foret, Paul, 151–52
Gilio-Whitaker, Dina, 28
Giorgio (protagonist). *See La Mucama de Omicunlé* (Indiana)
Gissibl, Bernhard, 119
Giuliani, Gaia, 60, 69
Glasgow COP 26 summit, 144
Glissant, Édouard, 49
Global North, hazardous waste from, 176; depicting unequal North-South relations, 165–70; overview, 161–62; question of "environmental racism," 162–65
Global South, 93; cinematic portrayal of toxic waste, 161–79; depicting unequal North-South relations, 165–70; overview, 1–4; photographing aftermath of natural disasters in, 144–60; question of "environmental racism," 162–65; resisting and imagining from, 4–9
Godoy, Fernando, 52–53
Goicochea, Adriana Lía, 128
gold, 76–77, 84, 94–96; extracting, 101–7
"Gold" (chapter). *See Guerras del Interior* (Zárate)
Gómez-Barris, Macarena, 8, 94, 146
gothic, engagement with: caring for "monstrous" children, 129–30; ecogothic lens, 129; flexibility of gothic, 128; homaging Lovecraft, 130–34; nature/landscape as source of fear, 128–29; overview, 127–28, 140; questioning limits of monstrosity, 134–40; resource gothic, 129

Grande sertão: Veredas (Rosa), 111–15, 117, 120–21, 122n3
Grande Sertão Veredas National Park, 112–15
Gran Sabana (region), 111, 115–18, 120, 122n6
great acceleration (paradigm), 4
Great Derangement, The: Climate Change and the Unthinkable (Ghosh), 4–5, 44, 141n7
Groom, Amelia, 217
Grosvenor, Gilbert, 62
Gudynas, Eduardo, 45
Guerras del Interior (Zárate), 76; anthropology of ontologies, 80–83; decolonial scopes of Indigenous activisms, 83–87; first chapter of, 76; limits of culturalist explanations in, 77–79; overview, 88; second chapter of, 76; third chapter of, 76
Guiana Highlands, 122n6
Guzmán, Patricio, 50
Guzmán-Conejeros, Rodrigo, 128

Hale, Charles, 36n12
Haraway, Donna, 9, 157n7
Harvey, David, 218
hazardous waste. *See* dumping of toxic waste, cinematic portrayal of
Heffes, Gisela, 88, 163
Heise, Ursula, 83
Henríquez, José M. Marrero, 88
Herdmark, Ann-Kristin, 177n6
Herrera, Eduardo, 218
heuristic imagination, 225–27
"Hidden Emissions, The" (report), 144
Hispanic Caribbean, 4
Hispanic Ecocriticism (Henríquez), 88
Höhler, Sabine, 119
Horn, Eva, 58
horror literature. *See* gothic, engagement with
Huaiquimilla, Kütral Vargas, 27

imagining (from the South), 4–9
Imataca Forest Reserve, 102
imperial city, decline of, 64–67
Implication, force of, 246
Inca Empire, 64
Inca Land: Explorations in the Highland of Peru (Bingham), 59; postcatastrophic visions of La Sierra in, 67–70

India, 61, 161
Indigenous activisms, decolonial scopes of, 83–87
Indigenous cadaver, reclaiming, 32–34
"infrapolitics," 36n8
Integral Peruvian identity. *See* ruins, emotional impact of
Integrated Communication Device (ACI), 222. See also *(In) Verdades: Uma heroína negra mudará tudo* (Ain-Zaila)
(In) Verdades: Uma heroína negra mudará tudo (Ain-Zaila), 216; collection action in, 224–25
heuristic imagination, 225–27; postapocalyptic nation-building in, 220–23; trauma and collective action, 223–25; trauma in, 223–24
IPAM. *See* Amazon Environment Research Institute
iron, 95–96

Jasanoff, Sheila, 228
Jimenez, Jonathan, 58
Jonk. *See* Jimenez, Jonathan

Kalén, William Johansson, 162, 164
Kawésqar (group), 50–52
Keetley, Dawn, 129
Kim, Sang-Hyun, 228
kimche, 27–28, 31
Kirschbacher, Felix, 58
Kohn, Eduardo, 81
Kopenawa, Davi, 147–48, 156n4, 209
Kressner, Ilka, 7
kultrún (sacred drum), 31–32, 239, 241
Küme Mongen (good living), 36n10, 239–41
Kupper, Patrick, 119

La Federica (train), 173
Lagunas, Samuel, 183–84
La Infancia del Mundo (Nieva), 218
"La materia" (poem), 30–32
La Mucama de Omicunlé (Indiana), 183; Beautiful Soul Syndrome, 189; bringing Marxism into queer ecological equation of, 197–98; critiquing self-actualization and escapism in, 185–87; critiquing sovereignty

in, 191–96; crossing boundaries in, 185–87; depicting development in ironic tone, 188; diminishing homophobia in, 194–95; end of, 198; fishermen embodying evil in, 189–90; framing Nature as both objectifying and dramatic, 189; and narratives centering on Anthropocene, 184–85; overview of, 183; previous studies on, 183–84; reinforcing Nature/Culture binary in, 188; release of true creativity in, 195–96; self-proclaimed environmental projects, 190–91; sexual and gender politics of, 196–97; toward queer Marxist ecology in, 196–98; upper and lower classes in, 187–88

land, reimagining: ecological constitution, 41–56; Mapuche poetics, 25–40; post-catastrophic landscapes, 57–72

Lao, Fernanda García, 133

La Paz, Bolivia. See *Sealed Cargo* (film)

La Sierra, 71n1; emotional impact of ruins, 61–63; externalization of nature, 64–67; overview of, 57–61; postcatastrophic visions of *La Sierra*, 67–70

La Sierra peruana, characterizing. *See* Peruvian Highlands, characterizing

LasTesis, performance by, 41–42

La Tierra Dormida (Cádiz), 218

Latin America: archives of the planetary mine, 92–110; artistic interventions in Chile, 41–56 Brazilian Afrofuturism, 216–34; cinematic portrayal of the dumping of toxic waste, 161–79; contesting climate unthinkable in, 1–19; cosmopolitics of the image, 201–15; depictions of postcatastrophic landscapes, 57–72; ecological conflicts in Peru, 75–91; ethics in *La Mucama de Omicunlé*, 183–200; ethnographic intervention, 237–44; fiction writing and environmental conservation, 111–24; Mapuche poetics, 25–40; photographing aftermath of natural disasters, 144–60; poetic intervention, 245–56; resisting and imagining from the South, 4–9; Río de la Plata gothic, 127–43; understanding climate unthinkable, 9–13

Latour, Bruno, 42–43, 89n5

Lauretis, Teresa de, 184

Lauro, S. J., 131

Law 21633, 29. *See also* Chile

"legislative theater," development of, 43, 53

Lenci, López, 59

Lértora, Carlos, 52–53

Lienlaf, Leonel, 27

life, micropolitics of: death by *terricidio*, 32–34; notion of "AzMapu," 27–30; overview, 25–26; poetic portrayal of death, 30–32

"*Literaturas Post-autónomas*" (Post-autonomous Literatures) (Ludmer), 245

lithium, 94–96, 105

local populations, removal of. *See* Grande Sertão Veredas National Park

Loncomeo dance (male performance). See *choike pürrun* (Mapuche dance)

López-Pellisa, Teresa, 218

Lo que la mina te da, la mina te quita (film), 94; context for, 101–2; exposing chemical process of mining, 103–4; illuminating contradictions in energy transition, 105; interpreting mining cabin, 105; "magic act" of the illusory, 106–7; and materiality of extractivism, 102–3; mission of, 102; overview, 107–8; recognizing Earth temporality of, 104–5

Los caprichos (Goya), 192

Los pasos perdidos (Carpentier), 111, 115–18, 120–21

Lost City of Incas (Bingham), 63

Lovecraft, H. P., paying homage to. See *Bajo el agua negra* (Enríquez)

Ludmer, Josefina, 245

machi (shaman), 238

Machu Picchu, 59, 61–63

Mackey, Allison, 127

Madariaga-Caro, Montserrat, 14, 25

Mad Max (film series), 222

Maduro, Nicolás, 101

Maldonado-Torres, Nelson, 86

Malm, Andrew, 1

Management Plan (Grande Sertão Veredas National Park), 113–14

Manifesto of the Communist Party (Marx), 188

Manquepillán, Faumelisa, 26, 27, 29; and death by *terricidio*, 32–34; "La materia" (poem), 30–32

Mapu, nurturing corporeality of, 30–32
"Mapuche micropolitics of resistance," 27
Mapuche poetics: death by *terricidio*, 32–34; enacting diverse micropolitics of life, 27–30; overview, 25–26; poetic portrayal of death in, 30–32
Marcone, Jorge, 77, 81, 88, 164
marea roja (toxic algal bloom), 52–53
Mariana dam disaster, 149–50
Maripil, Joel, 46
Martínez-Hernández, Lina, 184
Marx, Karl, 99, 188
Marxism, 197, 199n1
Mauro (character). See *Mugre rosa* (Trías)
Mayobre, Esperanza, 94
mediator, term, 151–52
Meiller, Valeria, 133
Meli Witran Mapu, 31
Merchant, Paul R., 8, 41
mercury, 103
Merlinsky, Gabriela, 9, 94
Miguel, Yolanda Martínez-San, 53
Millán, Moira, 25, 240
Million, Dian, 36n8
Minas Gerais, Brazil, 149–50
minerals, extracting: copper, 96–101; gold, 101–7; overview, 92–96, 107–8
minimal events, search for, 206–7
mining, socioecological impacts of. See planetary mine, archives of
Mirzoeff, Nicholas, 146–47
misanthropic skepticism, 86
Mises, Ludwig von, 218
mist, Río de la Plata region and, 137–38
Mockus, Antanas, 43
Monash Climate Change Communication Research Hub, 153
Mondzain, Jean-Marie, 144
Monsters, Catastrophes and the Anthropocene (Giuliani), 69
"monstrous" children, caring for. See *Bajo el agua negra* (Enríquez); *Mugre rosa* (Trías)
Moore, Jason W., 6, 183, 185
Moraña, Mabel, 132
morochucos, 64–66
Morton, Timothy, 89n4, 183, 187, 189, 193
Mother Earth, murder of. See *terricidio*

Movimiento de Mujeres y Diversidades Indígenas por el Buen Vivir, 25
Mugre rosa (Trías), 127; adaptation, 139; biological maternity in, 135; confinement in, 138; critiquing use of technoscientific terminologies, 135–36; daily landscape of, 134–35; denouncing fundamental failure of care, 136; freedom from past attitudes, 137; mist and fog in, 137–38; moving beyond ecophobia, 136–37; nature/landscape as source of fear in, 128–29; physical connections, 138–40; questioning limits of monstrosity, 135; setting of, 134
Mülchi, Hans, 51
Muñoz, Andrea and Germán Gana, 47
Murra, John, 68
Mutis, Ana María, 7

Nahuelpán, Héctor, 27
Namatjira, Albert, 151–53, 154
national cultural heritage. See ghost gum trees, destruction of
National Geographic Magazine, 62
National Museum of Chile, 34
national parks: creation of, 111–12; preserving collective environmental memory, 115–18; referencing in fiction, 112–15; relationship between fiction and conservation, 119–20
"Natura" (Jonk), 58
nature: aestheticization of, 198; bifurcation of, 202–3; civilizing, 119–20; coloniality of, 85; commodification of, 16, 88, 93, 146; contesting definition of, 156n4; emotional impact of ruins lost in, 61–63; establishing as subject of rights, 48–49; exploitation of, 92–110; externalization of, 64–67, 70; as gothic source of fear, 128–29; intertwining with built environment, 58–60; literary representations of, 75–91, 111–24; photographing, 144–60; postcatastrophic visions of *La Sierra*, 67–70; raw nature, 120; rights of, 14, 45; ruins and embrace of, 61–67; structuring representation of, 245–56. See also climate unthinkable, contesting
Nayar, Pramod N., 61
negotiator, term, 151–52
Negri, Antonio, 155

Neimanis, Astrida, 47
neoliberal multiculturalism, 36n12
"networked consciousness," 217
Nieva, Michael, 218
Nixon, Rob, 148, 150, 161
nonhuman agency, visibility of, 140
Norperuvian Oil Pipeline, 76–77
Northern Australia, ghost gum trees in, 150–53
North Troy Environmental Justice Film Festival, 178n7
novelas de la selva (genre), 122n9. See also *Los pasos perdidos* (Carpentier)

Obando, Andrés Ernesto, 57
O'Connor, James, 183, 185, 190, 197–98
oil, extracting, 76
"Oil" (chapter). See *Guerras del Interior* (Zárate)
Oloff, Kerstin, 184
OMA. *See* Orinoco Mining Arc
"116 Nations Adopt Treaty on Toxic Waste, L.A. Times" (report), 177n4
ontologies, anthropology of, 80–83
Orinoco Mining Arc (OMA), 101–2
Oruro, 170, 173–74
otherness, expressing, 207–8
Others, virtual sociality of, 157n6
Ozymandias effect, 61, 66. *See also* ruins, emotional impact of

Painter, James, 57
Paisajes peruanos (Riva-Agüero), 59; aestheticizing Peruvian history in, 66; nature appearing in, 65; postcatastrophic visions of *La Sierra* in, 67–70
Parikka, Jussi, 95, 104
Paterito, Gabriela, 50
Perez, Craig Santos, 49
Pérez, Nohemí, 94
Peru, ecological conflicts in: anthropology of ontologies in, 80–83; decolonial scopes of Indigenous activisms, 83–87; limits of culturalist explanations, 77–79; overview, 75–77, 88
Peruvian Highlands, characterizing: emotional impact of ruins, 61–63; externalization of nature, 64–67; overview, 57–61; postcatastrophic visions of *La Sierra*, 67–70
Peruvian Yale Expedition, 64
Pettinaroli, Elizabeth, 7
Pichinao, Jimena, 28
Pinda, Adriana Paredes, 27
Pinochet, Augusto, 29, 36n11, 41, 225
Pizarro, Francisco, 84
"places *for* disaster," 60
"places *of* disaster," 60
planetarity, 10
planetary mine, archives of: deterritorialization of subsoil, 96–101; materiality of extraction, 101–7; overview, 92–96, 107–8
Plantationocene, 146–47, 155
Plaza de la Dignidad, 46
poetry, illuminating "the paths not seen" with, 2–3
Polígono (neighborhood), 165, 167–68
Politics of Nature, 42
"politics of subsistence," 36n8
possible futures, fixing monopoly on, 217–20
postapocalyptic nation-building, 220–23
post-autonomous thinking, 245–46
postcatastrophic landscapes, depicting: emotional impact of ruins, 61–63; externalization of nature, 64–67; overview, 57–61, 70–71; postcatastrophic visions of *La Sierra*, 67–70
Prado, Esteban, 131
Pratt, Mary Louise, 3
Progress of the Storm, The (Malm), 1
Project Cybersyn, 230n7
Promel (company), 165
Pruneda-Paz, Dolores, 128
"Psychic Goya," 191–92
Pugh, Jonathan, 50
Punter, David, 127
pyrocumulonimbus, 153

queer Marxist ecology, 196–98
"queer" mode, plurality of, 193
Quijano, Aníbal, 84

radical rentierism, 101
Rama, Ángel, 128, 224
Ramella, Juana, 131

Ramiro, Joca (character). See *Grande sertão: Veredas* (Rosa)
Rancière, Jacques, 5
raros, term, 128
rationality, term, 89n3
rational politics, 77
Red Desert (Veiga), 149
red line, visualizing. See Mariana dam disaster
reflexive consumerism, 188
Reid, Julian, 47
representation, challenges to/expanded forms of, 42–45
resistance (from the South), 4–9
resource gothic, 129
Revista Endémico (magazine), 48
(R) Evolução: Eu e a verdade somos o ponto final (Ain-Zaila), 216
Riobaldo (character). See *Grande sertão: Veredas* (Rosa)
Río de la Plata, gothic tropes and: caring for "monstrous" children, 129–30; ecogothic lens, 129; flexibility of gothic, 128; homaging Lovecraft, 130–34; nature/landscape as source of fear, 128–29; overview, 127–28, 140; questioning limits of monstrosity, 134–40; resource gothic, 129
Rio Doce, death of, 149–50
Riva-Agüero, José de la, 59–60, 64–65, 67–68
Robalinho, Roberto, 144
Roberts, Brian Russell, 49
Rogers, Charlotte, 116
Rolnik, Guattari and Suely, 100
Rolnik, Suely, 27
Romantic wilderness, 112
Roque (protagonist). See *La Mucama de Omicunlé* (Indiana)
Rosa, João Guimarães, 111–15
Rosales, Antulio, 101
Rosso, Ezequiel, 224
"Ruin, The" (Simmel), 66
ruins, emotional impact of, 61–63
"Ruins and the Embrace of Nature" (Dwyer), 61
Rupailaf, Roxana Miranda, 27

Sabana, Gran. See Carpentier, Alejo
Samarco Mining Company, 149–50
Santos-Granero, Fernando, 80
Saramago, Victoria, 111
scenes. See ghost gum trees, destruction of; Mariana dam disaster; wildfires (in Australia)
science fiction (SF), 216–18, 220, 224, 226, 228–30. See also Brazilian Afrofuturism, climate fiction and
Scott, James, 36n8
Sealed Cargo (film), 161–63, 177n5; addressing harsh reality of rural life in Bolivia, 174; Andean rituals in, 174–75; connecting environmental issues with questions of power and ethnicity, 172; context of, 170; difference from *Arica,* 170–71; highlighting conflict between expert/scientific knowledge and local/popular knowledge, 171–72; Indigenous guerrilla in, 174; label shifting in, 171; last scene of, 175–76; opening scene of, 170–71; overview, 176–77; plot of, 170; repeated panoramic landscapes in, 173; Train Graveyard in, 173–74
self-actualization, critiquing, 185–87
Selgas, Gianfranco, 1, 92
sensible, distribution of, 5
Serafini, Paula, 9
sertão, term, 113
Shagulian, Jasmin Belmar, 237
shared intelligence, science fiction as, 227–30
Shaviro, Steven, 229
Silent Spring (report), 177
Simmel, George, 66
Simpson, Leanne, 36n8
"sites of situated hope," unearthing, 9
Skellefteå (neighborhood), 165–68
slow-motion toxicity, 161
Smith, Neil, 66
social monstrification, 132
socioenvironmental conflict, 75, 77, 79
sociotechnical imaginaries, 228
soil, 31, 68, 103–4, 147, 154, 163, 167, 172
Sommer, Doris, 43
South America. See climate unthinkable, contesting
Southern Cone, 4
sovereignty, critiquing, 191–96
Specters of Marx (Derrida), 178n9

Speed, Shannon, 35n4
Spivak, Gayatri, 10
Stengers, Isabelle, 202, 205, 207–8, 211
Stephens, Michelle Ann, 49, 53
Sterzi, Eduardo, 149
Stoermer, Eugene F., 144
stone guests, 77
subalternity, culture as function of, 183–84
subclinical toxicity, 169
subjugated knowledge, insurrection of, 80
subsoil, deterritorialization of, 96–101. *See also* Chile
Sweden-Chile, invisible toxic link of. See *Arica* (film)

Tánana, estar listo para zarpar (documentary), 50–51
technoscientific terminologies, critiquing use of, 135–36
technosublime, term, 231n8
Tentacle. See *La Mucama de Omicunlé* (Indiana)
terricidio, 14, 35; classification of, 25–26; death as nurturing corporeality of Mapu, 30–32; and micropolitics of life, 27–30; reclaiming the dead from, 32–34
third space, term, 151–52
Todd, Métis Zoe, 79
Todd, Zoe, 184
Torre Telefónica, 46
toxic algal bloom. See *marea roja*
toxic colonialism, term, 177n4
toxic commons, 163
toxic continuum, identifying, 163
"Toxic Discourse" (Buell), 177
toxicity, ghostliness, 164–65
Toxic Playground. See *Blybarnen* (film)
toxic waste, dumping. See dumping of toxic waste, cinematic portrayal of
toxic waste colonialism, 171
Train Graveyard, 173–74
"trans-corporeality at sea," 47
trans-corporeal spaces, 163
trauma, climate fiction and, 223–25
trees: Australian wildfires of 2019/20, 153–55; ghost gum trees, 150–53; sentience of, 82–83

Trexler, Adam, 141n2, 216
Trías, Fernanda, 127–29, 133–40
Tsing, Anna, 60, 163
Tsing, Anna Lowenhaupt, 155–56, 167
Tuohy, William, 177n4

ül (Mapuche orality), 237–38
UNESCO World Heritage Site. *See* Canaima National Park
United Nations Environment Programme Convention, 171
unthinkable, term, 9–10
Uruguay, 2, 6
Uyuni, obsolete trains in, 173–74

Vargas, Margarita, 51–52
Vargas, Rodrigo Pino, 178n7
Vargas-Weiss, Julia, 162, 164–65
Vásquez, Ana, 29
Vázquez, Juan José Ponce, 193
vegetable kingdom. *See* Canaima National Park
Veiga, Bruno, 149–50
Venezuela, 2, 15–16, 93–94, 116–17; gold extraction in, 101–7
veredeiros, 114. *See also* Grande Sertão Veredas National Park
Vice, 154
Vich, Víctor, 65–66
Vicuña, Cecilia, 94
Viggiano, Alan, 122n3
Vilcashuamán, 59
"Vilcashuamán: Stories in Ruins" (Beasley-Murray), 59
Vilcashuamán (Vilcas), Peru, 64–67
virtual sociality, 157n6
Visión de América (Vision of America), 115–16
visuality, complexes of, 146–47
"Vitcos: The Last Inca Capital" (Bingham), 62–63

wanderer, term, 151–52
warming world, responding to: archives of the planetary mine, 92–110; artistic interventions in Chile, 41–56; Brazilian Afrofuturism, 216–34; cinematic portrayal of the dumping of toxic waste, 161–79; cosmo-

politics of the image, 201–15; depictions of postcatastrophic landscapes, 57–72; ecological conflicts in Peru, 75–91; ethics in *La Mucama de Omicunlé*, 183–200; ethnographic intervention, 237–44; fiction writing and environmental conservation, 111–24; Mapuche poetics, 25–40; overview, 1–4; photographing aftermath of natural disasters, 144–60; poetic intervention, 245–56; resisting and imagining from the South, 4–9; Río de la Plata gothic, 127–43; understanding climate unthinkable, 9–13

Wars of the Interior. See *Guerras del Interior* (Zárate)

water, 47–48, 97, 103–4, 128, 132–33, 139, 168, 213, 220–22

Water Code of 1981, 47–48

Watkins, Susan, 227

weichafe (figure), 32–34

Western gaze, decentering. *See also* ghost gum trees, destruction of

What the mine gives, the mine takes. See *Lo que la mina te da, la mina te quita* (film)

Whiteside, Kerry, 43–44

Whyte, Kyle Powys, 184

Wiedemann, Sebastian, 201

wildfires (in Australia), 153–55

Williams, Raymond, 133, 226–27

Willi Mapu (group), 27

wood, extracting, 76

"Wood" (chapter). See *Guerras del Interior* (Zárate)

World Bank, 161

World Health Organization, 164

Wynn-Sivils, Matthew, 129

xapiri, seeing, 147–48

Xapirimuu (film), 202, 207, 208–13

Yaghan (group), 50–51

Yanomami (Indigenous group), 13, 18, 147–48, 155–56, 156n3, 156n5, 201–2, 204, 208–12

Yellowstone National Park, 111–12

Zanelli, José Carlos Díaz, 75

Zárate, Joseph, 76–79, 81–83. See *Guerras del Interior* (Zárate)

zombie, 141n5

"Zombie Manifesto, A" (Lauro and Embry), 131

zombies. See *Bajo el agua negra* (Enríquez)

www.ingramcontent.com/pod-product-compliance
Lightning Source LLC
Chambersburg PA
CBHW030821230426
43667CB00008B/1325